THE LAST WORD 2

THE LAST WORD

2

NewScientist

Edited by Mick O'Hare

with illustrations by Spike Gerrell

OXFORD
UNIVERSITY PRESS

OXFORD
UNIVERSITY PRESS

Great Clarendon Street, Oxford ox2 6DP

Oxford University Press is a department of the University of Oxford.
It furthers the University's objective of excellence in research, scholarship,
and education by publishing worldwide in

Oxford New York

Athens Auckland Bangkok Bogotá Buenos Aires Calcutta
Cape Town Chennai Dar es Salaam Delhi Florence Hong Kong Istanbul
Karachi Kuala Lumpur Madrid Melbourne Mexico City Mumbai
Nairobi Paris São Paulo Singapore Taipei Tokyo Toronto Warsaw

with associated companies in Berlin Ibadan

Oxford is a registered trade mark of Oxford University Press
in the UK and in certain other countries

Published in the United States
by Oxford University Press Inc., New York

British Library Cataloguing in Publication Data

Data available

Library of Congress Cataloguing in Publication Data

Data available

ISBN 0-19-286204-9

1 3 5 7 9 10 8 6 4 2

Typeset in 9.5/12pt Meta Normal
by Graphicraft Limited, Hong Kong
Printed in Great Britain by
Cox & Wyman Ltd, Reading, Berkshire

CONTENTS

INTRODUCTION

Some people worry about quantum mechanics. Others prefer to wonder why there is a hole in the top of their parachute. Some people probe the cosmos looking for signs of extraterrestrial life. Others wonder what might happen to their beer in outer space. Some people want to identify every gene in the human body. Others simply wonder why silver paper hurts the fillings in their teeth.

For every big question that somebody asks, there are thousands of smaller ones. And, just as its predecessor did, this second volume of questions and answers from The Last Word page of the weekly science magazine *New Scientist*, celebrates the small questions, the ones everybody asks and the ones the magazine's readers continue to solve. Why do ants run around happily inside a functioning microwave oven? Does chocolate really cause spots? Just why do some people stick out their tongue when they are concentrating hard?

And then there are conkers. One reader wanted to know the best way to toughen his! This not only brought forth many theories on the best methods but prompted non-British readers to question exactly what a conker was in the first place . . . Well the answer is contained in the following pages and, if you want to produce a champion, get reading.

The first volume of The Last Word proved to be an astounding success, giving rise to this second volume two years later. It's almost seven years since the Last Word page was introduced to *New Scientist*'s readers and they have enthusiastically supported it ever since. More than 750 questions have been answered, nearly all by readers, and it is far and away the most popular section in the magazine. Each week more than 200 questions are asked, only a fraction of which can actually appear on the page, and responses to those questions fill mailbags, fax baskets and e-mail archives every day.

But that does not mean that more new questions and replies are unwelcome. The column thrives on the input of inquisitive and knowledgeable readers. *New Scientist*'s website at www.newscientist.com

contains the complete archive of Last Word questions and answers, plus a full list of unanswered questions. You can post new questions for the page on the site, e-mail them to lastword@newscientist.com or write to The Last Word at *New Scientist* Editorial, First Floor, 151 Wardour St, London, W1V 4BN. Many people who were inspired by reading the first volume of this book have gone on to appear in this second edition.

And the question most frequently asked by readers? That would be why is the sky blue—it is asked by at least one reader every week. Sadly for them, it's one that has been posed already. But if you are still intrigued, after reading all about conkers in this book, try getting hold of the first volume to discover the answer.

MICK O'HARE

PLANTS AND ANIMALS

BEELINE

Q *My girlfriend tells me it is impossible to explain how the bumblebee flies. Apparently it defies the laws of physics. Is this true?*
TORBJØRN SOLBAKKEN
NORWAY

A The infamous case of the flightless bumblebee is a classic example of carelessness with approximations. It stems from someone trying to apply a basic equation from aeronautics to the flight of the bee. The equation relates the thrust required for an object to fly to its mass and the surface area of its wings. In the case of the bee, this gives an extremely high value—a rate of work impossible for such a small animal. So the equation apparently 'proves' bees cannot fly.

However, the equation assumes stationary rather than flapping wings, making its use in this case misleading. Of course, if equations fail in physics there is always empirical observation—if a bee looks as if it is flying, it most probably is.
SIMON SCARLE
LONDON

BIOSPHERE

Q *Hypothetically (because otherwise my mum would get mad), if I were to put my brother in a perfectly sealed room, how much plant life would I need in that room in order to maintain a balance of oxygen and carbon dioxide such that both my brother and*

my beloved plants would continue
to live?
GENE HAN
IOWA

 To simplify matters, you could supply your
brother's meals through an airtight hatch. The
plants would then only need to provide his
oxygen. If he spent all his time eating and dozing,
he would need about 350 litres of oxygen per day
(the amount of oxygen in 1.7 cubic metres of air).
This much oxygen is produced in full sunlight
by typical vegetation covering a floor area of
between 5 and 20 square metres. Using the most
productive 'C4 plants' such as sugar cane, you
could reduce the area needed to 2.5 square
metres. Your brother would exhale 350 litres of
carbon dioxide per day, which would enable the
plants to grow with an increase of dry weight of
430 grams per day.

Now let's muddy the waters. If his windows plus
artificial lights supply 10 per cent of full sunlight,
multiply the required area of greenery by a factor
of 10. If the lights go out at night, double the
area—more in winter. Plants photosynthesize
during the day more rapidly than they respire at
night. Therefore, as a reasonable approximation,
you can neglect the extra oxygen that plants
consume at night.

If you don't intend to feed your brother,
but hope he will survive by eating the plants,
remember most material a plant synthesizes
is indigestible, so double the area again. The
inedible parts of the plants plus your brother's
faeces would need to be decomposed or burnt
to carbon dioxide to recycle the carbon they
contain. So, if your brother is a well-trained
plant physiologist, this ambitious biosphere

might need to be a plant-filled room about 20 metres square.

Basis of calculations: daily energy requirement of an adult dozing is 1750 kilocalories per day. Energy content of 100 grams of sucrose is 400 kilocalories. Therefore an adult needs 1750/400 = 438 grams of sucrose per day = 1.28 moles of sucrose per day.

Respiration of this requires $1.28 \times 12 = 15.36$ moles of oxygen per day. (One mole occupies 22.4 litres, so this corresponds to $15.36 \times 22.4 = 344$ litres of pure oxygen per day.)

Photosynthesis rates of plants in the field under optimal lighting are between 10 and 30 micromoles (up to 70 in C4 plants) of carbon dioxide fixed per square metre per second (0.86 to 6.05 moles per square metre per day). For each mole of carbon dioxide the plants fix, they liberate a mole of oxygen.

Therefore the area required is somewhere between 18 square metres for the less productive plants down to 2.5 square metres for C4 plants.

Stephen Fry
Institute of Cell and Molecular Biology,
Edinburgh University

A This is an important question for our future in space. The first experiments with uncrewed ecosystems were performed by Russian scientists in the 1950s. This led to the crewed closed facility Bios-3 in 1965, a 315-cubic-metre habitat at the Institute of Biophysics in Krasnoyarsk, Siberia. *Chlorella* algae, which photosynthesize, were used to recycle air breathed by humans, absorbing carbon dioxide and replenishing it with oxygen. The algae were cultivated under artificial light. To achieve a balance of oxygen and carbon dioxide, one human needed 8 square metres of exposed

Chlorella. The algae tanks were stacked so they took less than 8 square metres of floor space. Water and nutrients were stored in advance— these were recycled too. By 1968, the system efficiency had reached 85 per cent by recycling water. Bios-3 has conducted tests with two and three people for up to six months.

NASA is also conducting Controlled Ecological Life Support Systems experiments. One of its core systems is the Biomass Production Chamber, a sealable steel chamber about 3.5 metres in diameter and 7.5 metres high with a plant-growing area of 20 square metres. In 1989, NASA completed BioHome, which integrated biogenerative components for recycling air, water and nutrients from human waste into a single habitat.

RUDY VAAS
BIETIGHEIM-BISSINGEN, GERMANY

CIDER BENDER

In autumn 1994, I planted a Cox's Orange Pippin tree alongside a Bramley cooker. In autumn 1995, each produced a small crop of their designated variety.

In autumn 1996, the Cox tree produced a few small Cox apples but the Bramley tree produced slightly sour Cox lookalikes— during apple-blossom time there had been frost and gales.

Can someone tell me what happened and what I can expect in autumn 1997? Please don't say Golden Delicious.

A. PERERA
BRISTOL, AVON

A Many apple trees (the Cox being one) are diploid (they have two sets of chromosomes) and will pollinate other nearby apple trees provided their flowering periods overlap. The Bramley is triploid (it has three sets of chromosomes) and produces very little usable pollen, so it is virtually ineffective as a pollinator. It therefore requires two other diploids for its own pollination and these also cross-pollinate themselves.

It could be that the Bramley's seedling quoted in this question is not actually a Bramley's seedling.

JAMES RICHARDS
CAMBRIDGE

CONQUERING CONKERS

Q *I was once advised by a friend that the way to strengthen a conker before a conker fight is to bake it. From my childhood I remember being told that the way to improve your conkers was to pickle them in vinegar. Which method produces winning conkers and why?*
BY E-MAIL, NO NAME OR ADDRESS SUPPLIED

A The simplest and best way to harden conkers is to put them away in a drawer until the following year. However, if they were not attached to a string when new and soft they will have to be drilled.

Both my children and grandchildren played with my leftovers, some of which were 50 years old. They have never been defeated.

F. GRISLEY
BARRY, SOUTH GLAMORGAN

A I have been in conker fights for about 50 years and I always soak them in vinegar. This hardens them into champion conkers. I was content with this method until a few years ago when I was beaten by someone who had smeared his conker with Oil of Ulay. Apparently, this made the conker more malleable, allowing it to absorb the impact of my prize pickled nut.

MICHAEL DUTTON
GLOUCESTER

A Neither baking nor pickling in vinegar is an effective way to strengthen conkers. Baking makes chestnuts brittle, which means they can be knocked off their string with a single blow. Pickling rots the inside. Varnishing, another technique used, is also ineffective (and readily detectable).

In fact, no intervention is necessary to toughen a conker. Simply avoid using chestnuts from the current season (conker fights are obviously an autumn pursuit) and use old ones instead. The older they are, the harder they are. Such conkers are readily identified—instead of having a glossy chestnut brown skin, they will look dull and dark, perhaps even black. And finally, make the hole for threading the string as narrow as possible.

NICK AITCHISON
LONGLEVENS, GLOUCESTERSHIRE

A The best ways to make a conker invincible are either to leave it for a year and use it in the season after you found it or, to speed up the process, bake it. Put all your conkers in the oven at Gas Mark 1 (120 °C) for about two hours.

Do not leave them any longer otherwise the flesh inside the conker will become charred and weak. Even if the heat breaks the shell of your conker, the flesh will be rock hard.

Do not put your conkers in vinegar. Although vinegar hardens the shells, the flesh will soften up if there is any gap in the shell, making the conker useless.

PATRICK WIGG
LONDON

Every conkerer disagrees with every other on how best to produce an invincible nut. As such disputes are an essential part of the sport, we leave the question with the totally contradictory answers given above—Ed.

A I was intrigued by all the dialogue about conkers, and I gather that they are chestnuts, somehow attached on a string. What is it you British do with the chestnuts? I suspect it doesn't involve consumption.

JENNIFER HOLTZMAN
NORTH HOLLYWOOD, CALIFORNIA

A What kind of game is conkers? Is it like a pillow fight with rock-hard chestnuts? Sometimes we yokels in the former colonies need a little edification . . .

JAY KANGEL
BY E-MAIL, NO ADDRESS SUPPLIED

Conker fighting appears to be a mainly British pursuit so we realize we have to enlighten our international readership. Conkers are the hard fruit of the horse chestnut tree. These are collected in autumn, removed from their spiky casing and left to mature. A hole is then drilled in the conker, and a string threaded through. The full rules of genuine competition are complex but the game as played by schoolchildren (and

overgrown schoolchildren) is between two opponents each with one conker. One player dangles their conker by the string, holding it steady, while the opponent swings their conker on its string and attempts to strike the hanging conker. Players take it in turns to do this until one conker is so damaged that it is dislodged from its string. The winner is obviously the player with the intact conker. Naturally, the stronger and harder the conker, the more chance of success.

This is perhaps further proof, if it were needed, of the British obsession for devising eccentric and meaningless methods of competition—Ed.

 The Australian game is called 'bullies' and is played in a manner similar to the way the English play conkers.

I know of it being played at least between 1900 and 1970 in western New South Wales and South Australia and before the advent of the TRS-80 and Commodore 64.

As most fair-dinkum Aussies with a bush background will know, the bully is the seed of the quandong tree. Also known as the wild peach, this widely distributed bush tree requires a host tree to survive and fruits annually, producing a tart, bright-red fruit, up to 40 millimetres in diameter.

This fruit has been an Australian delicacy in pies and jams for many years and is only now becoming more commercially available. The stone from the fruit is perfectly round and dimpled like a golf ball. It is usually about 20 millimetres in diameter, with the requisite internal nut and as hard as a stone. Drilling was a difficult task, with good bullies causing the demise of many a parent's drill bit.

Local rules governed the length of the string and the size of the playing circle. Ownership of losing bullies was not an issue, as my recollection is that losers shattered.

No heat treatment was needed or desired— fire and heat are necessary for the germination of most Australian bush seeds, so heating would certainly weaken the bully. I am unaware of an international challenge in this enthralling sport, but perhaps this could be proposed for the Sydney Olympics in 2000. My money would be on the bullies from the colonies.
JIM BILLS
BY E-MAIL, PROUDLY FROM AUSTRALIA

A My brother was disqualified from the school conker championships for having a conker that was vacuum-impregnated with epoxy resin.
J. MCINTYRE
BALSHAM, CAMBRIDGESHIRE

DAWN MISERY

Q *Why do roosters crow in the morning?*
BRENDA WARD
BY E-MAIL, NO ADDRESS SUPPLIED

A In fact, roosters crow throughout the day, although crowing tends to be more vigorous at dawn and decreases from late afternoon (about 3 p.m.). I have heard my roosters crow as late as around 5.30 p.m. (in summer). Crowing begins at or just before the crack of dawn—as my neighbours will testify.

This led me (with some persuasion from the local council) to soundproof and lightproof the coop. Apparently, both the light and the sound of wild birds' dawn chorus can stimulate crowing, which is a territorial response as well as a means

of showing off to females in the chicken run. Soundproofing and lightproofing are only partially successful, however, because the crowing seems programmed into the rooster's biological clock.

C. CROOK
PRESTON, LANCASHIRE

A The majority of birds take part in what is known as the dawn chorus. The singing is probably largely territorial. For about half an hour around sunrise, more birds will be singing than at any other time of the day. The average domestic rooster is considerably louder than other birds and is therefore the most noticeable at dawn.

However, roosters do not always crow at dawn. If your questioner had visited Cornwall on 11 August 1999, she would have been be able to hear the dawn chorus, including the roosters, during the false dawn at the end of the total eclipse of the Sun, which occurred during the middle of the day.

JON MILLER
HELSTON, CORNWALL

A Revenge for coq au vin.

SEAN KELLY
LEIGHTON BUZZARD, BEDFORDSHIRE

DESIGN FAULT

Q *While working in the garden, I saw a beetle walk past, take a wrong step and land on its back. Without my intervention it would have stayed in this position and probably died. Why is it that millions of years of evolution*

have not eradicated this basic and potentially lethal design fault?

GREG PARKER
BROCKENHURST, HAMPSHIRE

 If your correspondent had left the beetle in place on its back it probably would not have remained as it was until death. Beetles and other insects have a variety of mechanisms which they can use for righting themselves in these circumstances which, as the writer presumes correctly, must arise often and hazardously.

The most famous mechanism is used by the click beetles (*Elateridae*), which are able to launch themselves into the air by the sudden release of a blunt spine which is kept under pressure in a specialized groove on the venter.

As many readers will have noticed, the click beetle often makes several attempts before it lands on its feet but its success, given time, is assured.

Other less sophisticated beetle-correcting mechanisms include spreading the wings, reaching out with the legs, and rocking the body in a forward-aft or side-to-side motion.

Any readers wanting to learn more about the subject should see the following paper: 'Comparative righting behaviour of insects' J. T. Chao, Chung Hsing University, Taiwan, 1985.

CHRISTOPHER STARR
DEPARTMENT OF ZOOLOGY, UNIVERSITY OF THE WEST INDIES,
ST AUGUSTINE, TRINIDAD AND TOBAGO

 Only a minority of beetles possess a body plan that poses such a problem. For example, I have worked with several species of ladybeetle (*Coccinellidae*) in the laboratory, and most are able to right themselves with relative ease.

The species that do find themselves stranded on their backs tend to be the larger varieties that possess strongly convex elytra (the first pair of hardened, protective wings). Ladybeetles that do become stranded on a smooth surface will eventually unfold their membranous hind wings, which are normally hidden beneath the elytra, and then use these to right themselves. Part of the answer then, is that very few species become stranded and those that do eventually flip themselves over by means of their hind wings.

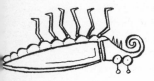

Over the long course of evolution it was probably quite rare for beetles developing in temperate forests and grasslands to encounter totally smooth surfaces or bare soil that was devoid of plant litter. Under normal circumstances, grass blades, fallen leaves and plant stems would offer a convenient hold for beetles that happened to become overturned.

The reduced rate of predation and numerous other benefits that are conferred by a hard protective covering, which far outweighs the occasional stranding, has contributed to the enormous evolutionary success of beetles. In terms of both absolute numbers and numbers of species, beetles are the most successful group of animals on the planet.

TOM LOWERY
PEST MANAGEMENT RESEARCH CENTRE,
ONTARIO

 I doubt whether the beetle was a healthy specimen that just happened to fall over and was unable to right itself. It is more likely that it was an old, sick or diseased specimen that was nearing the end of its life. When this happens in beetles, they lose a great deal of their mobility and coordination and they become very unstable when walking. They frequently fall over when placed on a hard, flat surface and are unable to right themselves.

I have observed this countless times in a number of beetle groups. In fact, while growing up I lived near Milwaukee, Wisconsin, in the US. We had a fairly large population of *Carabus nemoralis*, which is a ground beetle that was introduced from Europe into the US. I would frequently find beetles on the sidewalks on their backs. No matter how many times they were righted, they would invariably end up on their backs again, soon to die. I also observed beetles stagger out of the vegetation bordering the sidewalk, only to fall onto their backs. If these beetles were placed on their feet, even in the vegetation, they would stagger about and would fall onto their backs again when they encountered the sidewalk.

So, I suspect that the poor design is really a combination of dying beetles coupled with a smooth, hard surface—one that is not normally found in nature. Considering that roughly one out of every five living creatures is a beetle, and that they occupy virtually every niche and habitat known, I would suggest that beetles are, in fact, very well-designed animals.

DREW HILDEBRANDT
BY E-MAIL, NO ADDRESS SUPPLIED

EGGSACTLY

Q **Why are most eggs egg-shaped?**
MAX WIRTH
BOWNESS-ON-WINDEMERE, CUMBRIA

A Eggs are egg-shaped for several reasons. First, it enables them to fit more snugly together in the nest, with smaller air spaces between them. This reduces heat loss and allows best use of the nest space. Second, if the egg rolls, it will roll in a circular path around the pointed end. This means that on a flat (or flattish surface) there should be no danger of the egg rolling off, or out of the nest. Third, an egg shape is more comfortable for the bird while it is laying (assuming that the rounded end emerges first) than a sphere or a cylinder.

Finally, the most important reason is that hens' eggs are the ideal shape for fitting into egg cups and the egg holders on the fridge door. No other shape would do.

ALISON WOODHOUSE
BROMLEY, KENT

A Most eggs are egg-shaped (ovoid) because an egg with corners or edges would be structurally weaker, besides being distinctly uncomfortable to lay. The strongest shape would be a sphere, but spherical eggs will roll away and this would be unfortunate, especially for birds that nest on cliffs. Most eggs will roll in a curved path, coming to rest with the sharper end pointing uphill. There is in fact a noticeable tendency for the eggs of cliff-nesting birds to deviate more from the spherical, and thus roll in a tighter arc.

JOHN EWAN
WARGRAVE, BERKSHIRE

A Eggs are egg-shaped as a consequence of the egg-laying process in birds. The egg is passed along the oviduct by peristalsis—the muscles of the oviduct, which are arranged as a series of rings, alternately relax in front of the egg and contract behind it.

At the start of its passage down the oviduct, the egg is soft-shelled and spherical. The forces of contraction on the rear part of the egg, with the rings of muscle becoming progressively smaller, deform that end from a hemisphere into a cone shape, whereas the relaxing muscles maintain the near hemispherical shape of the front part. As the shell calcifies, the shape becomes fixed, in contrast to the soft-shelled eggs of reptiles which can resume their spherical shape after emerging.

Advantages in terms of packing in the nest and in the limitation of rolling might play a role in selecting individuals which lay more extremely ovoid eggs (assuming the tendency is inherited) but the shape is an inevitable consequence of the egg-laying process rather than evolutionary selection pressure.

A. MACDIARMID-GORDON
SALE, CHESHIRE

GLASS MENAGERIE

Q *I have just seen a film on television showing a variety of beautiful translucent deep-sea creatures, including a squid that was completely translucent except for its eyes and ink gland.*

Why are these creatures translucent, how is this achieved and, in the event that a gene for translucency exists, would it be possible (if not desirable) to make humans translucent?

J. FRIBERG
GOTHENBURG, SWEDEN

A Several multicellular organisms are almost completely transparent, and many possess at least some translucent body parts or tissues. The most striking animal examples include some deep-sea squids and the amphipod *Phronima*, several species of marine and freshwater shrimps, practically all 100 or so species of arrowworms (chaetognaths), the wings of some butterflies (*Callitaera menander*), the predacious aquatic larvae of the insect Chaoborus and even some species of fish, such as the catfish Kryptopterus.

Obviously, being transparent makes visual recognition by both predators and prey more difficult. In marine animals, transparency permits vertical migrations across water layers of differing hues and light intensities, without the animal having to worry about adjusting body colour. Even mammals possess transparent tissues such as fingernails and the lens of the eye.

A South Atlantic transparent medusa I picked up in 1967 was so perfectly transparent and biconvex that, by shining sunlight through it like a hand lens, I was able to light a cigarette with it.

Transparent tissues share some general features: no blood vessels or very few, an absence of pigment cells, extracellular spaces that are smaller than the wavelength of light, and a relatively regular and repetitive structural unit. Commonly mucopolysaccharides and collagen are involved in animal transparency, but glycoproteins (in jellyfish) and chitin (in insects) may also be found.

Some tissues are impossible to make translucent: nerves will always look white even in transparent organisms because of their high lipid content, and the retina has to remain pigmented because of its visual purple (rhodopsin).

Obviously, stomach contents cannot be made invisible.

The maintenance of transparency requires energy. Dead tissues eventually lose their transparency, a process that is accelerated by heat—look at the lenses of boiled fish eyes.

This all explains why the 'invisible man' of TV fame was never shown drinking a glass of wine.
V. MEYER-ROCHOW
OULU, FINLAND

LUCKY MARK

Q **Local birds tend to eat little black insects. So how come they void themselves on me from a great height with a white and annoyingly conspicuous product?**
M. ROGERS
GREAT HOCKHAM, NORFOLK

A It is a common misconception that the white droppings birds produce are faeces. In fact, they are urine. Birds excrete uric acid rather than urea because it is an insoluble solid. This way they avoid wasting water when urinating, and is just one of their adaptations for a good power–to–weight ratio.
GUY COX
UNIVERSITY OF SYDNEY, NEW SOUTH WALES

A The white material that comprises the droppings of birds, and indeed many reptiles, is their urine.

The more primitive vertebrates excrete toxic nitrogenous waste relatively directly, having masses of water at their disposal with which they can dilute substances such as ammonia.

However, birds and reptiles—at least lizards and snakes, with whose droppings I'm very familiar—are different. It would appear that the conversion of their toxic nitrogenous waste products into a relatively insoluble one that can then be formed into a paste was an evolutionary adaptation. This enabled them to lead a terrestrial rather than aquatic life, and even to live in ecological niches where water is scarce.

In such niches it is particularly important not to have to find extra water with which to dilute toxic waste products and flush them from the system, so birds and lizards solved this by evolving to produce a paste of insoluble and relatively nontoxic uric acid.

Interestingly, birds that consume a lot of roughage in their diet, such as the heather-eating grouse and ptarmigan, produce droppings that are very similar to guinea-pig faeces. Only here and there among the droppings is it possible to make out the telltale white patches of their urine, so copious is their production of faeces.

PHILIP GODDARD
BY E-MAIL, NO ADDRESS SUPPLIED

They do so from a great height because from a lower height it's just too easy to hit the target—no challenge at all. The deposit needs to be white so that, from said great height, they can see where it lands and who it hits.

S. B. TAYLOR
CANTERBURY, KENT

Your previous correspondents omit one fact, oviparity. The evolution of insoluble excreta has nothing to do with a 'good power–to–weight ratio' or the ability to 'live in ecological niches where water is scarce'.

It evolved because all birds and many reptiles begin their life inside an egg. Even heavy egg-laying amniotes that live in water as adults, such as penguins and crocodiles, must survive this early phase without poisoning their shelled enclosure with any water-soluble metabolites.

ÖRNÓFLUR THORLACIUS
REYKJAVIK, ICELAND

PEPPER POT

Q *This question was first asked on the newsgroup sci.botany, but nobody answered it. Are* New Scientist *readers able to help?*

When a pepper is cut open there is a space inside, but there are no gaps in the pepper where air could get through. What is the composition of gases in this space, and how did they get there? If a green pepper contains chloroplasts would there be more oxygen and less carbon dioxide in a green pepper than in a red, yellow or orange one?

ROSA CLEMENTS
HARROGATE, NORTH YORKSHIRE

A The questioner is not quite right in suggesting that 'there are no gaps in the pepper where air could get through'. Like most other plant surfaces the surface of a pepper or capsicum has stomata. These are orifices which are controlled by a pair of special cells, the guard cells, to open or shut as the plant requires.

They communicate with an extensive network of air spaces within the tissues, without which the gas exchange required for both photosynthesis and respiration could not take place.

The source of the air is, therefore, the atmosphere, via the stomata and the intercellular air spaces in the wall of the fruit. All capsicum fruit are initially green, with functional chloroplasts, so it is possible that there could be some enrichment in oxygen from photosynthesis at this stage, but not very much, because without gas exchange with the outside air there would be no source of carbon dioxide for further photosynthesis.

When a capsicum ripens to a red or yellow colour the chloroplasts cease to function, and turn into chromoplasts containing fibrous deposits of carotenoids and protein. At this stage photosynthesis has ceased and the internal gas will be unlikely to differ very much from the outside air other than in water vapour content.

Guy Cox
Sydney, New South Wales

 As the pepper develops, the gases from the atmosphere diffuse into the capsicum's growing cavity. The composition of the internal gases will depend on the respiration rate of the pepper as well as the resistance to gas diffusion. Generally, the more immature the pepper, the higher the respiration rate of the tissues.

We decided to test the internal gas composition of different coloured peppers in our laboratory using gas chromatography. The mean percentage levels of oxygen and carbon dioxide respectively were: green (19.85, 0.068), yellow (18.45, 1.08), red (18.36, 1.15).

It is possible that the higher oxygen/lower carbon dioxide levels in the green fruit were due to photosynthesis because the lab bench was in bright sunshine during all the measurements.

Normally, however, internal light levels are much too low to support any significant photosynthesis in harvested produce.

JULIA AKED AND ALLEN HILTON
SILSOE, BEDFORDSHIRE

POP THE QUESTION

Q *What is the difference between corn, sweetcorn and popping corn? What makes sweetcorn sweet and popping corn pop?*

ANDREA COWAN
ST. ANDREWS, FIFE

A Corn is the generic term to describe the fruit (grain) of cereal plants in particular. In Britain, corn generally refers to wheat. However, in the US, corn refers to maize, *Zea mays*. Popping corn and sweetcorn are just two of the many varieties of maize grown commercially. Each variety has different properties and is grown for different reasons.

The bulk of tissue within a grain of corn is called the endosperm. Endosperm is specialized storage tissue providing nutrients for the embryo when the seed germinates. It is also a source of carbohydrate for humans. In popcorn, *Z. mays* var. everta, the outer part of the endosperm is hard but the centre is soft. When the corn is heated the water in the central part turns to steam causing the seed to burst (making the pop that you hear) and turn inside out.

Sweetcorn, *Z. mays* var. saccharata, contains more sucrose in the endosperm than other varieties, which is what makes it sweet. It is

grown in the US and Europe as a vegetable, as well as in Mexico and South Africa where it is used in brewing beer.
ELIZABETH STRIPP
EXETER, DEVON

A Maize, like all cereals, stores a carbohydrate food reserve in its seeds. Sugars enter the developing seed and are converted to starch. As the grain matures, excess water is removed leaving a hard, dry starch.

In the maize varieties used for most purposes the sugar is all converted to dry starch: this type of corn is called flint corn. In sweetcorn the process is not completed by the time the plant is harvested, leaving the grain moist and sweet. In popcorn the centre of the endosperm is still rich in water at the time of harvest, but the outside is hard. On heating, the water in the centre turns to steam and blows the grain open.

There are other forms of maize including flour corn where the starch remains soft—this was used by Native Americans because it is easy to grind—and waxy corn which, on milling, produces a flour with the texture of tapioca.
JOHN GOODIER
LONDON

A Corn is simply the American word for maize. No other crop has received such intensive genetic study, and hundreds of different varieties are grown throughout the world. These are classified into five principal commercial types according to the structure of the kernel: dent maize, flint maize, flour maize, popcorn, and sweetcorn. These crops were cultivated in the Americas before the arrival of Europeans, maize being the basic food plant of all pre-Colombian American civilizations.

The grains of sweetcorn contain a glossy, sweetish endosperm which is translucent when immature. A recessive gene on the fourth chromosome prevents the conversion of some of the sugar into starch. It is harvested when young and immature, when the kernels are plump but still soft and milky. As soon as the silk threads above the winter husk wither and turn brown, the cobs can be broken off, boiled, and eaten as a vegetable. However, 20 minutes after picking, the sugar begins converting to starch. Therefore, corn on the cob should be absolutely fresh if you want it to be sweet. For the same reason canned or frozen sweetcorn is processed as soon as possible after harvesting. Canned sweetcorn is America's most popular vegetable.

The grains of popcorn are small with a high proportion of very hard endosperm and a little soft starch in the centre. On heating, the moisture in the centre expands as steam, causing the grain to pop and explode, the endosperm becoming everted as a palatable and fluffy mass. It can then be coated with freshly prepared butterscotch to make candied popcorn. Other types of corn will crack but will not explode. Popcorn is not grown in Britain, but is imported from the USA and, more recently, from South Africa.

GABRIEL LEVINE
TORQUAY, DEVON

RED OR WHITE?

Q *Why is red meat red and white meat white? What is the difference between the various*

animals that makes their flesh differently coloured?

TOM WHITELEY
BATH, SOMERSET

 Red meat is red because the muscle fibres which make up the bulk of the meat contain a high content of myoglobin and mitochondria, which are coloured red. Myoglobin, a protein similar to haemoglobin in red blood cells, acts as a store for oxygen within the muscle fibres.

Mitochondria are organelles within cells which use oxygen to manufacture the compound ATP which supplies the energy for muscle contraction. The muscle fibres of white meat, by contrast, have a low content of myoglobin and mitochondria.

The difference in colour between the flesh of various animals is determined by the relative proportions of these two basic muscle fibre types. The fibres in red muscle fatigue slowly, whereas the fibres in white muscle fatigue rapidly. An active, fast-swimming fish such as the tuna has a high proportion of fatigue-resistant red muscle in its flesh, whereas a much less active fish such as the plaice has mostly white muscle.

TREVOR LEA
OXFORD

 The colour of meat is governed by the concentration of myoglobin in the muscle tissue which produces the brown colouring during cooking.

Chickens and turkeys are always assumed to have white meat, but free-range meat from these species (especially that from the legs) is brown. This is because birds reared in the open will exercise and become fitter than poultry grown in restrictive cages. The fitter the bird, the greater the ease of muscular respiration, and hence the increased myoglobin levels in the muscle tissue, making the meat browner.

All beef is brown because cattle are allowed to run around in fields all day, but pork (even free-range) is whiter because pigs are lazy.

T. FILTNESS
WINCHESTER, HAMPSHIRE

RIG DWELLERS

Q *I work on a North Sea oil platform and have often wondered what our numerous gull inhabitants do during severe weather. They certainly cannot sit it out on the water or glide above the storm—and I assume they don't fly ashore. Just what do they do?*
CHARLIE MCGREGOR
THE NORTH SEA

A Some birds can sit out storms at sea. They can feed, and their feathers maintain waterproofing and warmth. Puffins and razorbills rest on land in Britain but winter in the mid-Atlantic. Force 10 gales are not a problem: if a large wave crashes over the birds, they merely bob up on the other side.
DOM JOHNSON
ZOOLOGY DEPARTMENT, OXFORD UNIVERSITY

A Some birds can ride out a storm pretty well using the wind. However, they cannot stay in the fixed vicinity of a rig at such times, so you wouldn't see them during storms.
JON RICHFIELD
DENNESIG, SOUTH AFRICA

A Birds do fly inland. For 12 years I lived in Birmingham—about as far from the sea as you can get in England. In stormy weather, flocks of gulls would sit on our wall and roof and scavenge from our dustbins, while nearby rubbish tips also had seagulls by the thousand.

PHILIP MUDD
BROADSTONE, DORSET

These apparently conflicting answers are all partly correct because what happens depends on the birds' species, their location, and the weather conditions.

Some birds can sit out storms on the water surface. The auks, which include the puffins and guillemots, leave land at the end of the breeding season and are adapted to living at sea. During their moult, puffins temporarily lose the ability to fly and have little choice but to stay on the water. Fortunately, because they are diving birds, temporary submergence in rough weather is not going to trouble them. Other diving birds, including ocean-going penguin species in the Southern hemisphere, can also sit out storms.

Larger species may take to the air. Gannets (which nest on land but spend a large proportion of their lives over the ocean), fulmars, petrels, and the larger gull species are all likely to do this. The albatross may prefer high winds as it can travel huge distances in storms.

Smaller gulls are more likely to seek refuge on land. But it should be remembered that gulls have become so successful at exploiting human rubbish dumps that some inland colonies in Europe have effectively lost their links to the sea.

During large storms, any ocean bird species can accidentally end up being wrecked on land. In Britain, one of these species is Leach's Petrel which, although ocean-going, is often blown onto land in stormy weather—Ed.

With thanks for help from the Royal Society for the Protection of Birds.

WATER EVERYWHERE

Q *What do seals, or any other marine mammals, drink?*
LAURA RICHARDS
WELLINGTON, NEW ZEALAND

A Marine mammals do not actually drink. They get all their water from the fish that they eat. They become dehydrated very quickly if they are not feeding adequately and, when in captivity or care, they must be fed freshwater from a pipe or bottle if they are not eating.
A. MORRIS
LONDON

MYSTERIES AND ILLUSIONS

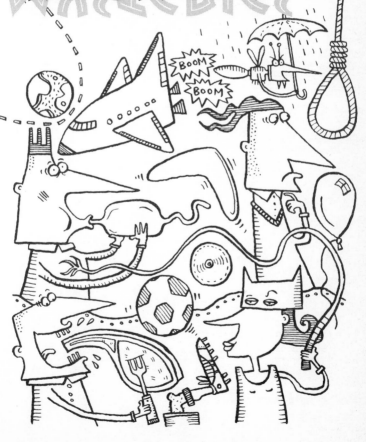

BOOM BOOM

Q *Why, when the space shuttle is returning to Earth, do you hear two separate sonic booms a few seconds apart?*
ANDY HORANIC
ORLANDO, FLORIDA

A There are two parts to the answer. First, why is there a sonic boom at all? The speed of sound in some media—air or water, for example—is the speed at which the molecules comprising the medium are able to pass along to their neighbours, via collisions, the information that something is approaching or receding.

When an object moves through such a medium at less than the speed of sound, the compression wave that it creates in the forward direction (as well as a corresponding rarefaction behind it) outstrips the motion of the object and, in general, the resulting local changes in density and pressure are quite modest.

When the object zips along faster than the message itself can travel, however, it can induce extremely abrupt and energetic pile-ups of molecules, which tend to get rid of their excess exuberance in a raucous manner.

While we normally can't see such a shock wave in the air, in water it is a familiar sight. It is simply the wake created by a boat that is moving faster than the waves on the surface of the water it is travelling through.

The water analogy also serves very well to illustrate why there should be two, or even more, sonic booms. We can see that a boat usually creates two separate disturbances, one that emanates from the bow and the other coming from the stern. These depart from the boat in two

V-shaped ridges. If there were such a thing as a 'supersonic submarine', the disturbances from bow and stern would, under conditions of full immersion, be cone-shaped because they can be expressed in three dimensions.

In the case of a supersonic aircraft, each of the many abrupt discontinuities in the airframe profile such as the wings or the tail fin also creates a series of these conical shock waves.

Like those created by a boat, the two dominant shocks from the space shuttle emerge from near the nose and from the tail, and they produce two sonic booms as they drag along the ground below the shuttle. To continue the water analogy, this phenomenon is, in effect, like a boat's wake meeting the shoreline.

MARK DAMASHEK
HAMPSTEAD, MARYLAND

As the space shuttle re-enters the atmosphere at supersonic speeds, it creates shock waves which produce sonic booms. Because the shuttle is large in size, at 37 metres long, a ground-based listener will hear two sonic booms. These are created by the nose and tail shock waves and occur about half a second apart.

Actually all supersonic aeroplanes including Concorde produce two sonic booms but, because they happen so close to each other, you hear them as a single sound.

REBECCA MARSHALL
WELLCOME WING PROJECT, SCIENCE MUSEUM,
LONDON

This is simply caused by the fact that when the orbiting space shuttle is re-entering the Earth's atmosphere at hypersonic speeds (Mach 15 and above) shock waves are generated from the nose of the aircraft (the first sonic boom to be heard) and also from the wing tips or the tail fin (the second sonic boom).

The delay between these two booms is caused by the distance between the nose and tail of the shuttle being magnified by the time the shocks reach ground level. The shock waves generated by the nose and tail start off a mere 37 metres apart, yet by the time they reach a listener on the ground this distance has increased significantly.

This magnification is caused by the slight difference in the angles of the shock waves on the nose and on the tail. By the time the waves reach the ground this tiny difference in angle has increased proportionately and that's why listeners can hear two booms.

Models of the space shuttle being tested in a NASA wind tunnel show shock waves clearly emanating from various points on the shuttle's surface, and these are caused by the numerous discontinuities that are found there.

RICHARD MARTIN
LONDON

€NERGY LOSS

Q *What is the so-called 'slingshot effect' used to accelerate interplanetary spacecraft? It obviously makes use of the gravitational attraction of a planet, but my naïve understanding of physics tells me that any kinetic energy gained on approaching a body would be lost as potential energy on leaving. How does the spacecraft extract energy from the planet?*
DAVID BATES
ELY, CAMBRIDGESHIRE

I had the same problem as your questioner when I first heard about Voyager using the 'slingshot effect'.

Clearly a probe will not make any net gain in energy by simply falling through a stationary gravity field.

However, Jupiter and its gravity field are moving around the Sun at a speed of about 1300 metres per second and any probe passing behind the planet will be accelerated by this moving gravity field much as a surfer is pushed forward by a wave. The energy does not come from the gravitational field but from the kinetic energy of the moving planet which is slowed by the tiniest amount in its orbit, causing it to drop ever so slightly closer to the Sun.

The planet speeds up as it falls towards the Sun, and paradoxically it ends up moving more quickly than it did before. Moving Jupiter closer to the Sun by 10^{-15} metres (about the diameter of a proton) would yield more than 416 megajoules.
Mike Brown
Knutsford, Cheshire

GREEN HAM COMMON

Q *What causes the greenish iridescent sheen that I often notice on bacon and ham? Is it harmful, and why does it vanish when the product is heated? Does this occur on any other foodstuffs?*
Georgina Godby
Cambridge

A You are likely to find such a sheen on foods containing traces of fat in water. When it is cool this mix separates out microscopically into a film, like oil on a wet road.

In some types of cold meats, such as sliced silverside of beef or some hams, you may see a handsome opalescence. The beauty of an opal results from light being refracted and diffracted by arrays of microscopic beads of glassy material in a matrix of a different refractive index. In the meat, the effect is caused by microscopic spheres of fat dispersed in watery muscle tissue. Heat up the meat and you destroy the droplets and change the optical character of the matrix so that the effect is spoilt.

JON RICHFIELD
DENNESIG, SOUTH AFRICA

 The green colour that is sometimes observed on bacon and ham is the result of the action of non-pathogenic bacteria which break down the oxygen transport protein myoglobin to produce porphyrin derivatives. These derivatives are large heterocyclic compounds which can have greenish colours.

STEPHANIE BURTON
DEPARTMENT OF BIOCHEMISTRY AND
MICROBIOLOGY, RHODES UNIVERSITY,
SOUTH AFRICA

 My father, working alone in the Australian bush in the 1920s and 1930s, ate meat either fresh, soon after it was killed, or after it had been hung in a tree long enough for it to turn a brilliant green. The meat was put into a bag to keep the flies off it.

He claimed that the green colour showed that the meat was no longer dangerous to consume, and it certainly never killed him. However, there is little doubt that it did change the flavour considerably.

JAN MORTON
WEST LAUNCESTON, TASMANIA

Iridescence is caused by light striking a surface and being scattered. The scattered waves interfere to produce a spectrum of colours which changes depending on the position of the observer. However, if you see a bright green colour rather than a mere iridescent sheen then your meat may be only for the hardy stomachs of those who tramp the Australian bush—Ed.

HAPPY RETURNS

Q *Why do boomerangs come back?*
ADAM LONGLEY
BARRY, SOUTH GLAMORGAN

A A boomerang is like two spinning aeroplane wings joined in the middle. It is held almost vertically before it is thrown end over end. Because it spins in this way, the top wing actually goes away from you faster than the bottom wing. This makes the sideways push on the top wing (similar to lift on an aeroplane wing) stronger than that on the bottom wing, so the boomerang gets tilted over, just as you would be if someone pushed on your shoulder, and its flight pattern begins to curve.

Similarly, if you ride a bicycle and lean over, the bicycle will turn, eventually going in a circle. The boomerang does too.
ALAN CHESTER
SHEFFIELD, SOUTH YORKSHIRE

A Returning boomerangs work by a combination of aerodynamic and gyroscopic effects. A boomerang is essentially a rotating wing with two or more aerofoil-shaped blades. It is thrown with its plane of rotation at about 20 degrees to the vertical and so that it spins rapidly (typically about

10 revolutions per second) with the uppermost blades travelling in the direction of overall motion. Therefore, the blade at the top moves through the air faster than the lower one. The faster-moving blades generate more lift than the slower-moving ones. This produces an overall force in the direction of turn, plus an overturning torque.

The rotation of the boomerang makes it behave like a gyroscope. When the overturning torque occurs, the gyroscopic effect makes the boomerang turn (or precess) about a different, near-vertical, axis. This continuously changes the boomerang's plane of rotation, causing it to travel around an arc back to the thrower.

Other effects are also evident in the boomerang's motion, such as its tendency to lie flat as it returns to the thrower—its plane changes from 20 degrees from the vertical initially, to horizontal at return. This is caused by a number of aerodynamic effects combined again with gyroscopic precession. The most significant effect is that the blades on the leading side of the rotating boomerang generate more lift force than the blades on the trailing side, because of the disturbed airflow on the trailing side. This again causes rotation which leads the boomerang to spin towards the horizontal plane. An article by Felix Hess in the November 1968 edition of *Scientific American* explains this process in detail.

RICHARD KELSO AND PHILIP CUTLER
UNIVERSITY OF ADELAIDE,
SOUTH AUSTRALIA

A The simple answer to this question is that most boomerangs don't come back and were never intended to do so. The Australian Aboriginal people made the boomerang for hunting and fighting rather than for sport or play, so they did not make the so-called returning boomerang over most of the Australian continent. For them, the real returns of boomerang-throwing came in the form of fresh food or the besting of an enemy.

I have seen the Warlpiri people throw a karli boomerang and hit a target at well over 100 metres. Particularly skilled users of the karli throw this deadly weapon with surprising ease. The Warlpiri also manufacture the wirlki (also known as the 'hooked' or 'Number 7' boomerang) which is used for fighting.

Across Australia, even in those areas where the boomerang is not made, there is near universal use of paired boomerangs as rhythm instruments in ceremonial contexts. Such boomerangs are still traded for ritual use across thousands of kilometres.

There are and have been an astonishing variety of boomerangs from Australia. For an accessible account see *Boomerang: Behind an Australian Icon* by Philip Jones, published by the South Australian Museum.

CHIPS MACKINOLTY
NIGHTCLIFF, NORTHERN TERRITORY

HOW ABOUT GNAT?

Q *How is it possible for gnats to fly in heavy rain without being knocked out of the air by raindrops?*
L. PELL
UFFINGTON, OXFORDSHIRE

A A falling drop of rain creates a tiny pressure wave ahead (below the raindrop). This wave pushes the gnat sideways and the drop misses it. Fly swatters are made from mesh or have holes on their surface

to reduce this pressure wave, otherwise flies
would escape most swats.
Alan Lee
Aylesbury, Buckinghamshire

 The world of the gnat is not like our own. Because
of the difference in scale, we can regard a collision
between a raindrop and a gnat as similar to that
between a car moving at the same speed as the
raindrop (speed does not scale) and a person
having only one thousandth the usual density—
for example, that of a thin rubber balloon of the
same size and shape. A balloon is easily bounced
out of the way, and would burst only if it was
crushed up against a wall.
Tom Nash
Sherborne, Dorset

IT'S A CRACKER

 Why does the end of a whip crack?
David Innes
Farnham, Surrey

 The crack is actually a sonic boom, caused by the
end of the whip breaking the sound barrier. This is
possible because a whip tapers from handle to tip.
When the whip is used, the energy imparted to the
handle sends a wave down the length of the whip.
As this wave travels down the tapering whip it
acts on a progressively smaller cross section and
a progressively smaller mass.

The energy of this wave is a function of mass
and velocity and since the energy of the wave
must be conserved, it follows that if mass is
decreasing as the wave travels down the whip,
then velocity must increase. Therefore, the wave
travels faster and faster, until by the time it

reaches the tip it has attained the speed
of sound.
MIKE CAPP
OXFORD

A When the wave reaches the tip of the whip its
energy must be dissipated. Some goes to the
air and some into a reflected wave that travels
back up the whip. At the point that the initial
wave reaches the tip and is about to embark
on its return it undergoes a brief but
enormous acceleration. The result is that
it moves supersonically.
ANDREW PLANT
LYMINGTON, HAMPSHIRE

LIGHT FANTASTIC

Q *Why do my CDs show weird patterns on
the metallic disc when viewed under the
fluorescent light of my bedroom? In daylight
or under a normal filament bulb, they just
look shiny with a slight ringed texture. The
pattern I see under the fluorescent light only
occurs on one side—the non-music side—
and is irregular.*
ROBIN SMITH
SWAYTHLING, HAMPSHIRE

A Fluorescent lights look white but they are not.
They cheat by having a hidden mix of distinct
colours which appears to our eyes as a close
approximation to white. Colour TVs and computer
screens show exactly the same white, but here it
is made up of just three colours.
  Objects viewed in fluorescent light may not
show their true colours because they can only

reflect the limited number of colours given off by the light source. Light sources that give off the full spectrum, such as the Sun, may be needed to show the object as it really is.

The grooves in the reflective coating in a CD cause interference between light waves. This interference is most distinct with monochromatic light, which has a single wavelength. Thus any source of light that does not give off a full spectrum will produce a more spectacular effect.

MIKE COON
MAIDENHEAD, BERKSHIRE

THE LIVING DEAD

Q *On her latest CD, the American artist Laurie Anderson uses the refrain 'Now that the living outnumber the dead . . .' Is this true? If so, when did it happen? If not, when might it happen, if ever? Do we have good estimates of population numbers before recorded history?*

JOHN WOODLEY
TOULOUSE, FRANCE

A The answer below is based on some calculations which were published by the International Statistical Institute.

If the world population had always been increasing at its present rate, doubling within an average human lifespan, then the living would indeed outnumber the dead.

However, this is not what has happened. There have been very long periods in the past when the population hardly grew at all, but when deaths continued to accumulate. For historical periods, there is a surprising amount of information on population figures, including censuses conducted by both the Romans and the Chinese.

Before then, there are estimates based on the area of the world which was under cultivation or used for hunting, and of the numbers of people who could be supported per acre using these methods of food production. According to estimates assembled by J-N. Biraben, the world population was about 500 000 in 40 000 BC. It grew—but not at a steady rate—to between 200 and 300 million in the first millennium AD, and reached 1 billion early in the 19th century.

On multiplying the population numbers by the estimated death rates, you discover that the total number of deaths between 40 000 BC and the present comes to something in the order of 60 billion. The present world population is still only about 6 billion.

Although no great claim can be made for the accuracy of the historical estimates, the errors can hardly be so large as to affect the conclusion that the living are far outnumbered by the dead. This has always been the case, and will continue to be so into the indefinite future.

ROGER THATCHER
NEW MALDEN, SURREY

In the Garden of Eden, the living (2) outnumbered the dead (0).
G. L. PAPAGEORGIOU
LEICESTER

In the Indian epic poem *Mahabharata the Eldest Pandava*, Yudisthira was asked many questions, including the one posed above, by the god Yama, who was the keeper of the Underworld and all that is righteous. This was to test

Yudisthira's knowledge, power of reasoning, and truthfulness.

Yama was disguised as a stork guarding a pond from which Yudisthira's four brothers drank. They were unable to answer a single question and were all struck dead. The stork Yama asked 'Who are the more numerous, the living or the dead?' Yudisthira answered 'The living, because the dead are no more!'

Yama accepted this and all the other answers given by Yudisthira and, with great pleasure because Yudisthira was actually the son of Yama, blessed him and revived all of his dead brothers.

SHAFI AHMED
LONDON

TACKLING INFLATION

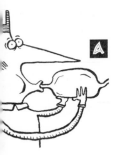

Q *Why are long balloons much harder for me to blow up than round balloons?*
RORY SULLIVAN
EVESHAM, WORCESTERSHIRE

A Inflating a balloon involves creating excess pressure. The rubber provides the equivalent of a bubble's surface tension. The excess pressure in a bubble is inversely proportional to its radius, so it is much more difficult to blow up a balloon of a smaller radius.

(A good party trick is to connect two balloons of different sizes by a short, closed pipe. Asked what will happen when the tube is opened, many people say that the air will flow from the larger balloon into the smaller one, equalizing the sizes. In fact, the exact opposite is true.)

Also, the force that must be overcome to inflate a balloon is proportional to the surface

area—which is at a minimum for any sphere. This is why bubbles are spherical.

RICHARD SMITH
MELKSHAM, WILTSHIRE

The difficulty of blowing up long balloons is caused by the extra pressure required when blowing up any balloon of small diameter in relation to a balloon of larger diameter. It requires a set amount of stress (engineers define stress as force per unit area using units such as kilograms per square centimetre) to cause the balloon rubber to stretch.

The stress in the rubber is proportional to the amount of air pressure, and to the ratio in balloon diameter to the thickness of the rubber.

For two balloons which have the same rubber thickness, the smaller one will require more pressure to produce the same stress. In fact, if a small balloon's uninflated diameter is a half of that of an uninflated large balloon, it will take twice the air pressure to inflate it.

Therefore, long balloons are much harder to blow up than round balloons because the long balloons have a much smaller diameter than the round ones. This effect also comes into play when you are blowing up round balloons of different sizes.

It also explains why a balloon is easier to blow up once the rubber starts to stretch, when it gets thinner. This decreases the pressure required to continue inflation. And, as the balloon inflates, the resulting larger diameter also lowers the required pressure.

As a rule, it takes less and less pressure as the balloon grows to continue inflation. But shortly before the balloon bursts, the required pressure usually increases noticeably. This is because, just before rupture, the rubber requires additional stress to expand further.

BARRY SPLETZER
BY E-MAIL, NO ADDRESS SUPPLIED

ROPE TRICK

Q *Under what circumstances are ropes and braids actually stronger than the same number of individual fibres and why?*
PATRICK ANDREWS
CAMBRIDGE

A Ropes, braids and woven fabric can sometimes exhibit strength greater than the individual strengths of the same number of fibres combined in any particular direction. When fibres are tested for tensile strength, weak spots along the length of the fibre will tend to form weak links—these are the areas where the fibre will break.

But in a rope, the weak spots in the fibres become randomized along the length of the rope. The twist in the rope increases fibre-to-fibre pressure perpendicular to the axial rope direction. The resulting increase in friction enables adjacent fibres without weak spots in the same area to add strength to each other. Normally, optimum strength in a rope is achieved when the twists in the fibres' single-thread, ply and cabled construction are axially oriented in the direction of the rope axis. Too little twist will not develop adequate compressive force and resulting friction to achieve optimum strength. Too much twist will place the fibres' single-threads, ply or cable in a state of shear and this too will reduce the rope's strength.

In the case of some braids and woven fabrics, the interlacing of yarns causes compression where one perpendicular or angular yarn crosses a yarn in the opposite direction. With braids and woven fabrics for a given size and type of yarn, there is an optimum construction that will maximize yarn-to-yarn compression with resulting frictional properties that will achieve the greatest strength. Less yarn will give insufficient compressive force, friction and resulting strength properties. However, if too many fibres are involved, shear will be increased and the yarn will be weaker.

BOB WAGNER
NORRISTOWN, PENNSYLVANIA

 A rope is stronger than a parallel group of individual fibres because of the 'twist'. This twist ensures that, during loading, any eccentricity in the load (bending as opposed to pure tension) is very quickly dissipated. The axis of a fibre on the outer surface of a bend (and hence under higher stress) moves to the inner surface within half a twist along the rope, and vice versa. All the filaments therefore carry a broadly equal share of the stress.

When a similar group of parallel filaments is loaded, any eccentricity in the load will ensure that some filaments are under a greater stress than others. They will break first, and the load they had been carrying will be thrown onto the remaining filaments; if any of these are also close to their breaking point then these will break and so on, starting a tearing effect.

The effect is very real: the tensile strength of individual cement-reinforcement glass filaments is about 3.5 gigapascals while the strength of a

strand of 204 parallel filaments is only about half
of this.

PHIL PURNELL
MATERIALS RESEARCH GROUP,
ASTON UNIVERSITY

SHEDDING LIGHT

Q *During a physics A level practical lesson, my
tutor placed a lit candle on a turntable. When
the table revolved we expected to see the
flame on the candle point outwards but,
instead, it pointed inwards. The school head
of science couldn't explain this. Can anyone
else?*

RUTH HAVELAND
BETWS-Y-COED, GWYNEDD

**Yes they can but, despite a large number of
replies, we found we had to knit many together
to get a clear picture. First of all, there was a big
problem—Ed.**

A My first reaction to the problem was not to believe
it. I tried the experiment and, sure enough, it
didn't behave as described. The flame trailed
behind the candle as it orbited the centre of the
turntable, just as it trails behind as you walk
along with the candle.

GARETH KELLY
HEAD OF PHYSICS, PENGLAIS SCHOOL,
ABERYSTWYTH, DYFED

A One of the great pleasures in life is lying in bed on
Sunday morning catching up with the latest *New
Scientist*, starting, perversely, with The Last Word.

This weekend was, however, different. First light found me in the kitchen with a candle on a rotating cheeseboard. At a speed of approximately 60 revolutions per minute the flame simply trailed behind the candle, showing no tendency to move out or in. I repeated the experiment later in the day on a gramophone turntable at 78 rpm, with the same result. Am I missing something?

JOHN ASHTON
MONMOUTH, GWENT

Yes, men of Dyfed and Gwent, you are missing something, although we commend your industry and integrity. So first of all . . . —Ed.

To see this effect, the candle must be effectively enclosed, otherwise it streams backwards. So, candle in jam jar, jam jar on edge of turntable.

DAVID MAY
PHYSICS TEACHER,
HIND LEYS COMMUNITY COLLEGE,
SHEPSHED, LEICESTERSHIRE

The reason the candle flame points inwards is the rotating table sets up a weak centrifuge.

DAVID BLAKE
STIRLING

As the air in the jam jar is being spun in a centrifuge, the denser air moves out with predictable consequences—Ed.

The candle flame bends towards the inside of the turntable for the same reason that flames move up rather than down. The heated gas of the flame is less dense than the cooler surrounding air, and the denser surrounding air moves out, forcing the candle flame in.

If I were to get really picky, I would argue that the less dense candle flame is accelerated more by the same centripetal force. Newton's law says that for the same force, the product of mass and acceleration is the same. So if the mass is smaller the acceleration must be more. At school level, it's simpler to think that the force has more effect on the denser air.

Sue Ann Bowling
University of Alaska,
Fairbanks

You can also think in terms of reference frames or do the maths—Ed.

 Understanding why the candle flame points inwards is made easier by considering a similar problem in a linear reference frame. Driving in your car which contains a helium balloon held by a string, you brake hard. What happens to the balloon? While you slam forward against the seat belt, the balloon goes towards the back of the car. This is because the air in the car has inertia and continues forward just as you do, and the balloon reacts by floating towards the lowest pressure, lowest density portion of the air mass at the back of the car.

Similarly, the candle flame is buoyant, its shape resulting from a complex interaction between the hot burning wax at the wick and the heating of the surrounding air. So, the flame also floats in the direction of lowest pressure— towards the axis of rotation. To complete the comparison, the candle, like the car, is accelerated with respect to the air surrounding the flame, so the air is moving radially outwards relative to the candle. The flame reacts by floating inwards.

Tom Trull
University of Tasmania

A The air in a closed container would displace the less dense gases in the flame towards the centre of rotation under the centripetal force field. The flame will make an angle arctan (a/g) with the vertical (where a is the centripetal acceleration).

This effect is demonstrated by a helium-filled balloon in a car. The balloon leans forward under acceleration, backwards when braking and towards the inside of bends. The same formula applies. For a car rounding a curve of 20 metres' radius at 50 kilometres per hour the lean should be about 44 degrees.
NEIL HENRIKSEN, RECTOR,
THE JAMES YOUNG HIGH SCHOOL,
EDINBURGH

And a simpler demonstration of the same effect—Ed.

A If you place a spirit level on the turntable pointing away from the centre like a bicycle wheel spoke, and rotate, the bubble quickly moves inwards. The more massive spirit has pushed the lighter bubble there.
COLIN SIDDONS
BRADFORD, WEST YORKSHIRE

TEN-BOB SWERVER

Q *I am well aware (having played many ball sports) of the Magnus effect which causes a ball that is spinning clockwise (when viewed from above) to swerve towards the right. Similarly, a ball struck with backspin will travel with a long, floating flight. These effects can be seen with leather footballs,* *tennis balls and table tennis balls. However, when applying spin to one of those plastic footballs sold at petrol stations and on beaches, the opposite is observed: clockwise spin produces a right to left swerve, and back-spin produces a viciously dipping shot. These balls are really only larger versions of table tennis balls, and similarly devoid of dimples and surface markings, so why should their responses to spin be opposite?*

RICHARD BRIDGEWATER
WALSALL, WEST MIDLANDS

This phenomenon was dealt with in some detail in 'The seamy side of swing bowling' (*New Scientist*, 21 August 1993, p. 21) and is best explained in terms of 'boundary-layer separation'.

When a ball travels through the air its surface is covered by a thin coating of air that is dragged along with it. Beyond this lies undisturbed air. Between the dragged air and undisturbed air lies a thin boundary layer. At the front of the ball, this layer moves slowly. But as it travels round the ball, it speeds up and exerts less pressure (as dictated by Bernoulli's law, which states that the faster a fluid flows, the less pressure it exerts).

At some point, the boundary layer separates from the ball's surface. If the ball is smooth and not spinning, this happens at the same point all round the ball. But if the ball is spinning, the boundary layer separates asymmetrically, so the boundary layer covers a larger area on one side than on the other. The result is a larger region of low pressure on one side of the ball than the other, which pushes the ball to one side.

In a conventional swing (produced by the Magnus-Robins effect), the spin of the ball carries a very thin layer of air along with it. This pushes the point of boundary-layer separation towards the back on the side of the ball where the spin is moving in the same direction as the surrounding airstream, and towards the front on the side that is moving against the air stream. The result is lower pressure on the side where the boundary layer has become extended, which causes the ball to swing in that direction. That's why a clockwise spin causes the ball to move from left to right. (Another way of describing what happens is to say that the shift in the point of boundary-layer separation pushes the flow lines of the air round the ball—the ball's wake—to one side, so that the ball swerves to the other.)

All this assumes that the flow in the boundary layer is laminar, with smooth tiers of air on top of each other. In practice, part of the airflow may be turbulent, with air moving chaotically throughout the layer, and this is where reverse swing can occur. Experiments show that turbulent layers stick to the surface of the ball longer than laminar layers. So if the boundary layer is turbulent on one side and laminar on the other, the pressure will be lower on the turbulent side and the ball will swing to that side.

Under certain circumstances, turbulence can develop first on the side of the ball which is moving against the airstream, so that the boundary layer here separates later. The result is a reverse swing. Whether turbulence will develop depends on the type of ball, its speed, size, and spin, so reverse swing is seen more commonly in some sports than others (see the following answers).

Sports such as cricket, which use balls with seams, give bowlers additional opportunities to produce both swing and reverse swing through turbulence. Skilful players can bowl so that the ball spins with its stitched seam always facing at a particular angle to the oncoming air. The seam affects the airflow, making the boundary layer turbulent on only the seam side of the ball. The boundary layer thus separates later on this side of the ball and the result is a vicious swing.

Bowl fast enough and that swing can be made to reverse. At the very high speeds produced by world-class bowlers (more than 130 kilometres per hour), the air moves so fast that the boundary layer becomes turbulent even before it reaches the seam of the ball. In this case the seam pushes the boundary layer away, encouraging it to separate from the ball earlier on the seam side. The ball then unexpectedly swerves in the opposite direction from usual. This is the notorious ten-bob swerver.

The effect can be produced by ordinary cricketers too, if their ball is scuffed, as a rough surface allows a turbulent boundary layer to develop more easily. Deliberate scuffing is, of course, against the rules—Ed.

 The reverse swerve body on a plastic football is due to boundary-layer separation. On the side of the football where the relative velocity of the air and football is larger, the flow in the boundary layer becomes turbulent. On the other side it remains laminar. The laminar boundary layer separates from the ball's surface once the airstream is no longer pushing it onto the surface. By contrast, the turbulent boundary layer remains in contact with the surface farther round the ball. This results in the wake behind the ball being deflected in the opposite direction to the rotation of the ball. And it produces a force towards the side of the ball that is moving in the opposite direction to the airstream (from right to left for a ball spinning clockwise).

Experiments show that the main factor governing the direction in which a ball swerves is the ratio of the rotational speed of its surface to the ball's translational speed. The reverse swerve occurs when this ratio is small (less than 0.4), while the Magnus effect occurs at higher ratios, which probably explains why the faster-spinning tennis ball swings in the opposite direction to the football.

OLIVER HARLEN
UNIVERSITY OF LEEDS

 The swerve of a spinning ball is commonly ascribed to the Magnus effect but, more than a century before Heinrich Magnus, Benjamin Robins studied spinning cannon balls and in 1742 he published a description of why, even on windless days, they swerved off-course.

BRIAN WILKINS
WELLINGTON, NEW ZEALAND

Many publications do now refer to the Magnus-Robins effect. It is perhaps worth remembering that Isaac Newton commented in 1672 on how the flight of a ball was affected by spin—Ed.

STRANGE NATURE

CLOUDING THE ISSUE

Q *Why do clouds darken to a very deep grey just before it is about to rain or prior to a heavy thunderstorm?*
MATT BOURKE
GRACEVILLE, QUEENSLAND

A Clouds darken from a pleasant fluffy white just before rain begins to fall because they absorb more light then.

Clouds normally appear white when the light which strikes them is scattered by the small ice or water particles from which they are composed. However, as the size of these ice and water particles increases—as it does just before clouds begin to deposit rain—this scattering of light is increasingly replaced by absorption.

As a result, much less light reaches the observer on the ground below and the clouds look darker.
KEITH APPLEYARD
DUNDEE, TAYSIDE

FLAT STONED

Q *While walking down a stony beach and skimming some stones into the sea, I noticed that the flattest pebbles (and the best ones to use for skimming) tended to accumulate at the top of the beach, furthest from the sea. Why is this?*
MATTHEW RICHARDSON
EDINBURGH

The amusing answer that all of the flat stones are found in this location because those nearest to the sea have already been used up by people

**skimming earlier was provided by an astonishing
25 people. Perhaps there is a ring of truth in their
suggestions. However, we suspect that the true
reasons are given below — Ed.**

A The coastal sediment system is dynamic. Wave
action constantly washes the sediment along the
coast until it reaches a point where the wave
energy is reduced by the local topography to the
extent that transport of the sediment is no longer
possible, and it is deposited.

Prevailing weather conditions, the aspect of
the beach and the coastline, wave fetch, offshore
topography and so on, cause some beaches, over
time, to be subjected to a higher wave energy
regime than others. Consequently, sediments are
sorted by their location or graded by their ease of
transport, and beaches which are subjected to a
higher wave energy regime are constructed largely
from those sediments which are more difficult to
move and vice versa. In a similar manner, areas
that are further up a beach will have a much
higher energy regime than areas nearer the water,
because only the more energetic waves are
capable of reaching them.

Normally, the sediments which are more difficult
to move by wave action are those which are larger
and heavier or those which are more spherical
and have a smoother surface, as these have a
relatively low surface–to–weight ratio. However,
most stones above a threshold size, certainly
including those large enough to be held in the
hand for skimming, are not picked up entirely by
most waves, and are instead rolled to a new
location. The shape of flat stones resists this
rolling action and therefore they are more
resistant to movement by waves, despite their
large surface area relative to their weight.

Furthermore, when taken as a group rather than individually, stones with flat surfaces display characteristics of increased resistance to wave energy. On any shingle beach it can be observed that they sometimes tend to stack like coins and wedge between each other. This grouping makes them much more difficult for a wave to move and explains their location in an area of the beach that has a higher energy regime. Of course, their deposition in groups of this nature would also make them more noticeable and more enticing as a supply of skimming stones.

MARK ENGLAND
KIDDERMINSTER, WORCESTERSHIRE

A Sediment that is transported by waves running up a beach tends to reflect the contrast in power between the swash and the backwash. Because the wave's energy is dissipated as it runs up the beach, and water tends to be lost by percolation into the beach sediment, the backwash is weaker than the swash. The swash is often able to carry coarse and fine particles up the beach, but the backwash is only capable of carrying the fine particles, leaving the coarser particles stranded at the top of the beach.

This crude sorting process usually reflects relatively high magnitude, low frequency, storm events. The waves associated with these events leave behind 'lag' deposits of coarse sediment which tend to stay put until they can be shifted by the next event of a comparable, or higher magnitude. The 'armoured' beds associated with mountain rivers commonly display a similar accumulation of coarse sediment.

However, it is not just a particle's size which determines the ease of its transport, and hence the sorting of sediments, but also its shape. The weight of a particle, and the friction that it generates by its contact with the layer below it, determine its resistance to flowing water. And its cross-sectional area that is exposed to the flow determines the magnitude of the fluid force that is required to move it from its resting position.

Flat particles within a flow of water will tend to rotate to lie flat. This gives them a high friction contact with the surface on which are resting and a relatively small cross section in relation to their weight exposed to the flow. Because of this, they tend to be less mobile than a sphere of equal weight. Therefore, when a mixture of particles of different shapes but uniform weight is present, flat particles, once they are in position at the top of a beach, will tend to remain there for the greatest time.

NICK SPEDDING
DEPARTMENT OF GEOGRAPHY, UNIVERSITY OF ABERDEEN

FLOATERS

Q *What is the force that drives an isolated and floating piece of wheat or rice breakfast cereal through the milk to the side of the bowl where it aggregates with its companions?*
JOHN CHAPMAN
PERTH, WESTERN AUSTRALIA

A The force is due to an imbalance in the pull from the surface tension of the liquid around the sides of the floating piece of cereal. A simple experiment explains what is happening.

You need tap water and two polystyrene cups plus two small pieces taken from a third cup (two 1-centimetre diameter circles will do nicely). Fill the first cup up to within 1 centimetre of the rim, fill the second cup to the top and then carefully add more to the second cup until the water is up over the top of the cup but not spilling over, that is, until the water is held in a convex bulge above the top of the cup by surface tension.

Now place the small circles of polystyrene in the middle of each. The piece floating in the partially filled cup will, with a little prompting, move to the side of the cup and be held there. By contrast, the piece floating on the convex bulge of the water in the second cup will remain near the centre. Furthermore, if you push the piece to the edge of the cup, say with the tip of a pencil, the edge repels the small piece towards the centre with considerable force.

This is all caused by the surface tension of the water. In the partially filled cup, the water surface curves up to meet the polystyrene. This occurs because water molecules are more attracted to polystyrene than to each other. The water forms the convex bulge at the top of the second cup because the surface tension constrains the liquid surface to the smallest area possible, which similarly accounts for the spherical shape of liquid drops.

The water also curves up to meet the small circle of polystyrene on all sides. Where the water meets the polystyrene of the small circle, the surface tension pulls on each contact point in a down and outward direction provided by the angle of contact with the water. When the circle is in the middle of the cup the pull on the circle on one side is directly balanced by

the pull on the opposite side, because the
water curves up to meet the circle equally
at all points.

However, if the piece is moved towards the side
of the partially filled cup, the upward curve of the
water surface near the cup side reduces the curve
of the surface in contact with the circle. This
increases the outward pull on the side of the
circle nearest to the cup edge, resulting in a net
force towards the side of the cup.

The effect also accounts for the clumping
together of cereal pieces on the surface of milk
in your bowl and similar behaviour of leaves
and twigs on ponds and lakes.

RAY HALL
WARRENVILLE, ILLINOIS

Maybe it is a defensive strategy, they huddle
together like bison, to protect each other from
the predator (you). Or maybe it's just the surface
tension in the milk.

PER THULIN
NO ADDRESS SUPPLIED

The fact that rice and wheat—or any grain for that
matter—can gravitate towards its companions in
this fashion, depends upon their being able to
'sense' their way towards the common centre
of mass. This ability is known as the grain of
common sense.

Research has shown that when human beings
are dropped into a large bowl of milk this flocking
or aggregation potential is entirely lacking, thus
proving that they don't have a single grain of
common sense.

MARTIN MILLEN
KIDLINGTON, OXFORDSHIRE

GAS GASSING

Why does speaking through helium raise the frequency of the sounds emitted, even when the final transmission to the hearer is through air?
DAVID BOLTON
MOSGIEL, NEW ZEALAND

A Sound travels faster in helium than in air because helium atoms (atomic mass 4) are lighter than nitrogen and oxygen molecules (molecular mass 14 and 16 respectively). In the voice, as in all wind instruments, the sound is produced as a standing wave in a column of gas, normally air. A sound wave's frequency multiplied by its wavelength is equal to the speed of sound. The wavelength is fixed by the shape of the mouth, nose and throat so, if the speed of sound increases, the frequency must do the same. Once sound leaves the mouth its frequency is fixed, so the sound arrives to you at the same pitch as it left the speaker. Imagine a rollercoaster ride. The car speeds up and slows down as it goes around the track, but all cars follow exactly the same pattern. If one sets out every 30 seconds, they will reach the end at the same rate, whatever happens in between.

In stringed instruments, the pitch depends on the length, thickness and tension of the string, so the instrument is unaffected by the composition of the air. Releasing helium in the middle of an orchestra would therefore create havoc. The wind and brass would rise in pitch, while the pitch of the strings and percussion would remain more or less the same.

In the *Song of the White Horse* by David Belford, the lead soprano is required to breathe in helium to reach the extremely high top note.
EOIN MCAULEY
DUBLIN

GREENER AND GREENER

Q *When waste paper is recycled the new product is clearly of a poorer quality than the original paper. Is there a further deterioration if recycled paper is itself recycled?*

Can papers recycled several times be separated from non-recycled waste paper? And is there a limit to the number of cycles that constituent materials can go through?
MICHAEL GHIRELLI
HILLESDEN, BUCKINGHAMSHIRE

A Paper-making fibres, which usually originate from wood pulp, can be used about six times. Each time they are used the fibres deteriorate—they wear down and become shorter. Because of this it is necessary to add virgin fibres to strengthen and maintain the quality of the finished paper. However, some papers require special character-istics or are used for purposes best met by virgin fibre alone.

In the normal course of sorting domestic waste paper, it is practically impossible to identify and separate those papers made from virgin fibre from those which contain recycled fibre. It may be easier to separate commercial waste paper, but only if the supplying source (such as a printer) knows exactly what the paper is made of and keeps it totally separate from any other waste paper.

RUTH MACKMAN
PULP AND PAPER INFORMATION CENTRE,
SWINDON, WILTSHIRE

A Recycling causes paper to deteriorate in two ways. First, the cellulose fibres which give the paper its strength are shortened, so the resultant paper is dense, stiff, less opaque, less durable and has low resistance to tearing. Secondly, impurities, mainly printer's ink, make the paper less white. If the same paper is recycled again and again, both these problems will increase each time.

The number of cycles that paper can go through is limited by the fact that it has to support its own weight during the manufacturing process, otherwise it tears and production time is wasted. The mills avoid this by adding virgin pulp, which is perfectly legal. The National Association of Paper Merchants allows its 'Approved Recycled' logo to be used on paper which contains only 75 per cent 'genuine waste fibre'.

Paper that is to be recycled is separated into two main categories: pre-consumer waste (from the mill itself, for example) and post-consumer waste (the 'real' recycling we do at home and in the office).

Paper can be recycled into products such as packaging materials without having to be de-inked, or it can be put through a de-inking process and used to make low-grade materials such as newsprint or paper for directories. This sort of recycling is universally acknowledged to be a 'good thing'.

However, top-quality recycled paper, which might be used for company stationery where colour consistency is required contains up to 25 per cent virgin pulp, and up to 65 per cent pre-consumer waste. Only 10 per cent comes from post-consumer waste. Even then the post-consumer waste has to be bleached after de-inking.

In this way the paper in top-quality products can be recycled indefinitely. But many people are horrified when they discover the very low post-consumer waste content of such paper, and the additional bleaching means that the overall benefit to the environment is questionable.

JONATHAN COOLEY
EGHAM, SURREY

 Recycling shortens and weakens the cellulose fibres in paper until they break. The resulting pulp has to be used in progressively lower-grade applications.

When paper gets to the bottom of the reuse chain, in a product such as corrugated cardboard, a different method of recycling is needed. You can fill an old cardboard box with corrugated card, urinate on it and leave it to decompose to compost.

Then you can use the compost to mulch trees and regenerate the cellulose fibres this way. Food boxes such as cereal packets can also be composted in a garden compost heap. They provide air spaces, aerating the compost until they rot down. Card adds useful 'roughage' to your compost heap, offsetting the excess nitrogen and general sliminess that results from too many grass cuttings.

The Centre for Alternative Technology in Machynlleth, Powys, has lots of information on

composting for anybody wanting to take this further.
DAVID EDGE
HATTON, DERBYSHIRE

GUST QUEST

Q **What mechanisms are responsible for causing the wind to blow in gusts?**
CHRIS LONG
SUSSEX

A Near the surface of the Earth, friction slows the wind. Turbulence is almost always created by layers of air moving at different velocities and this enhances or reduces the surface wind. The enhancements are the gusts. Strong turbulence is also created by obstructions such as buildings, which is why city centres are notoriously gusty.

If the surface is sufficiently warmer than the air above then convection will produce columns or walls of warm air called thermals. These will rise from the surface, and draw in currents of air to the base of the rising column. These currents can add to the mean wind to produce gusts that are longer lived that the usual turbulent gust.

In addition, if the convection is strong enough, it may produce shower clouds by condensation of moisture in the thermal as it rises and cools. Subsequent evaporation can then result in columns of cold air rapidly descending from these clouds to produce violent gusts at the surface. These are sometimes called squalls.
MIKE BRETTLE
CARDINGTON, BEDFORDSHIRE

MERCURY RISING

Q *On a recent flight, I was studying a card listing items that were prohibited by airlines. I was amazed to see that I couldn't take a mercury thermometer on a flight. Why on earth not?*
RICK ERAHO
CLECKHEATON, WEST YORKSHIRE

A Planes are largely made from aluminium and, surprisingly, a very small amount of mercury can destroy a large amount of aluminium. Despite its apparently inert behaviour, aluminium is actually a rather reactive metal which will combine violently with oxygen in air. However, this reaction quickly produces a thin, tough oxide layer which stops further attack. The process of anodizing the aluminium thickens this layer to give better protection.

Mercury has the ability to disrupt this protective oxide layer, and the results can be spectacular. It can dissolve aluminium to form an amalgam which may break up the oxide layer from below—presumably the initial attack occurs through tiny faults in the oxide.

Many years ago a technician working for me spilled a few drops of mercury on his wooden bench, which had heavy aluminium angles screwed round the edges to protect it. Next morning large holes were eaten through the aluminium, the wood nearby was deeply charred, and large fragile towers of friable aluminium oxide had grown like strange corals.

This used to provide a fine chemistry experiment but it is now frowned upon because of the toxicity of the mercury.

On one occasion a passenger in front of me was prevented from carrying a barometer onto an aircraft because it was on the list of prohibited articles, even though this particular barometer was empty. With difficulty I persuaded the staff that it was harmless. They did not realize it was the mercury that was dangerous, they thought it was just barometers *per se*. I wonder what they thought an altimeter measures . . .

HARVEY RUTT
DEPARTMENT OF ELECTRONICS AND COMPUTER
SCIENCE, UNIVERSITY OF SOUTHAMPTON

Given the mobility of liquid mercury, the corrosive amalgam may form deep within the structure. An aircraft in which mercury has been spilled must be put into quarantine until the amalgam makes its presence known. Ultimately, the aircraft is likely to be scrapped because the engineering textbooks state that the amalgam slowly spreads like wood rot to adjacent areas.

ROD PARIS
AIR MEDICAL LIMITED,
OXFORD AIRPORT, KIDLINGTON, OXFORDSHIRE

Mercury, along with many other common chemicals, is classified under 'dangerous goods' in international regulations developed by the International Civil Aviation Organization, which is part of the UN. You are not permitted to carry this substance, or any article containing it, aboard an aircraft in hand luggage or checked-in baggage. An exception is made for small clinical thermometers in protective cases for personal use.

Should mercury-containing articles need to be transported they must be consigned as airfreight. The ICAO rules specify in detail how this must be done.

Don't think that you can afford to ignore these restrictions. In Britain, endangering an aircraft by taking aboard dangerous goods could result in a charge and hefty fine under the 1982 Civil Aviation Act. In the event of a mercury spillage the aircraft would need to be taken out of service. The airline and/or its manufacturer may try to recover costs from you or your employer.

JAMES HOOKHAM
FREIGHT TRANSPORT ASSOCIATION,
TUNBRIDGE WELLS, KENT

NOT MIDSUMMER

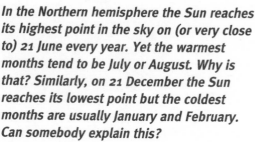

Q *In the Northern hemisphere the Sun reaches its highest point in the sky on (or very close to) 21 June every year. Yet the warmest months tend to be July or August. Why is that? Similarly, on 21 December the Sun reaches its lowest point but the coldest months are usually January and February. Can somebody explain this?*

WOLFGANG WILD
GRANADA, SPAIN

A The Earth has a certain heat capacity, which leads to a thermal time lag. Therefore, when a hemisphere is experiencing its longest day, it is still warming up, and it will not reach its warmest until a few weeks later. Similarly, on the shortest day it is still cooling down and will not reach its coolest until a few weeks later.

AIDAN WESTWOOD AND STEPHEN COLLINS
UNIVERSITY OF LEEDS

A Although the Northern hemisphere receives most sunlight at the end of June, it is rather like heating up a room with a gas fire. Although the fire gets hot very quickly, it takes a while for the air in the room to heat up. The same applies to the atmosphere of the Earth. Likewise, the room does not get cold the instant the fire is turned off—the air gradually cools in the same way the air in the Earth's atmosphere does.

IAN HEDLEY
POOLE, DORSET

RED HOT

Q *What causes the colours that form on a clean iron or steel surface after it has been heated and cooled for tempering? The colours range from yellow when the metal was heated to about 200 °C, through gold, brown, purple, blue and finally black when heated to about 600 °C. And because the oxidized blue or purple finishes on steel mechanisms have often survived unmarked in clocks from the last century, what is the physical nature of this transparent and very durable coloured layer?*

JOHN ROWLAND
ALLESYREE, DERBYSHIRE

A The hot furnace gases that are used for heat-treating steel oxidize the alloying elements, such as chromium, to form a thin surface film. These surface films interfere with visible light waves to produce the colours that your correspondent mentions.

The thickness of the films determines the apparent colour of the steel as it interacts with light of different wavelengths. Thinner films, which

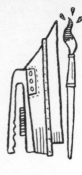

are formed at lower temperatures, seem yellow or gold. Thicker films make the steel appear light blue. The thickest films seem midnight blue and finally black.

Temper colours on clean, bare steel are actually quite fragile, and are quickly lost if rusting thickens the surface film by depositing layers of hydrated iron oxides. Many parts of the hundred-year-old clocks mentioned in the question owe the durability of their temper colours to the practice of dipping the tempered steel in sperm whale oil. The sperm oil gives a transparent, waxy protective covering to the oxide films, preserving their colours for posterity. Widespread use of this technique has had the obvious disadvantage of producing a serious shortage of sperm whales.

DALE MCINTYRE
DHAHRAN, SAUDI ARABIA

RIP OFF

Q **Why does newspaper always seem to have a preferential direction for tearing? It looks homogeneous to the naked eye. Other sorts of paper, such as A4 laser printer paper, seem homogeneous in appearance and in tearing.**
BOB JOHNSON
DURHAM

 Most paper is made on machines at high speed. The sheet is made by draining a dilute suspension of fibres and mineral fillers on a table of continuous and fast-moving synthetic sieve-like wire. The sheet is then consolidated by pressing and it is dried on heated cylinders before being reeled.

As the paper-making suspension is discharged from a vessel known as the flow box or head box onto the rapidly moving wire, most fibres—which are cylindrical in shape—become aligned in the direction of the wire's movement. This is called the 'machine direction'. The orientation of fibre in the paper structure allows the sheet to tear more easily in the machine direction than in the 'cross direction'—where the fibres are presented sideways.

The strength of all paper made this way is influenced to varying degrees by the directional properties of the sheet. The addition of fillers and mechanical or surface treatment during manufacture may help to modify or reduce this. Newsprint tears more easily in the machine direction than printing paper, because it contains mostly fibre and is lighter in weight.

R. I. S. GILL
LEEDS

RUSTLE RIDDLE

Q ***What generates the energy that makes thin, white supermarket bags so noisy?***
LUCY BIRKINSHAW
LEICESTER

A The energy is generated mostly by you, because the bag will not rustle by itself. The noise is caused by sharp movements of the kind you get when a stiff plate buckles or gets rubbed. The bags are made of polyethylene film which, untreated, is waxy and floppy and not very noisy. It is elastic rather than plastic, so it absorbs stresses quietly.

However, to make the bags, the film is stretched to get it thin enough to be convenient to handle and cheap enough to give away with the goods. This partly aligns its molecules into stiffer sheets. To make the bags look better and the contents more anonymous, manufacturers add fillers for colouring and further stiffening. The result is a bag which audibly protests every crinkle, crumple and abrasion.

JON RICHFIELD
DENNESIG, SOUTH AFRICA

WIND UP

Q *Why is it more windy in winter, when the input of solar energy is least?*
D. J. H. WORT
WHITBY, NORTH YORKSHIRE

A It is not the local input of solar energy that determines windiness, it is the differential in heating of the Earth's surface between cold polar regions and hot tropical regions. This leads to a circulation of air being set up to even out these differences, with general rising of hot air near the equator being replaced by air from near the poles.

If the Earth did not spin, this would lead to a wind at the Earth's surface blowing from the north in the Northern hemisphere and vice versa in the south. However, because of the Earth's rotation, the wind gradually curves to the west (this is called the Coriolis effect) and air cannot move directly from pole to equator in one step. The single convection cell splits into three, with rising air not only at the equator but also in the temperate regions, and subsiding air in the subtropical desert regions, as well as in the polar regions. Thus, there is a region of conflicting winds in the temperate regions (between approximately 40 and 60 degrees north and south).

In the Northern hemisphere temperate zone, cold northeast winds battle with warm southwest winds. The strength of these winds is less in summer because Arctic land masses are warmed by near continuous sunshine, and the difference in temperature across the temperate zone is not particularly marked. In winter, when the pole is in near continuous darkness, areas near it become very cold. Temperature differences across the temperate zone become large, strengthening vertical air currents in the convection cells, and making surface winds much stronger.

This is less noticeable in the Southern hemisphere. Southern polar regions are land covered by ice (Antarctica) or ocean, neither of which are warmed much in summer. This causes the temperature differential across the temperate zone to be much the same all year.

GRAHAM HUGHES
SUTTON, SURREY

 In winter, our weather is affected predominantly by depressions which are formed where warm, moist air, heated in the tropics, and cold, moist air from the North Pole, meet over the Atlantic. The warm air rises above the cold. Both air masses spiral upwards and anti-clockwise, forming a low pressure area, the energy being supplied by latent heat stored in water vapour in the air masses.

Air from surrounding higher pressure areas flows into the depression to reduce the pressure difference causing wind.

JOE WALTON
LONDON

Wind velocity in Britain is determined by the interaction of continental and north Atlantic air masses, rather than the solar heating of the islands. Strong winds are associated with the passage of small, deep depressions, which result from the mixing of warm Atlantic (Gulf Stream) and cold polar air. They drift eastwards with the prevailing Atlantic airflow.

As Europe cools in winter, air pressure rises and a macroscopic high forms over the continent. In the Northern hemisphere, air circulates clockwise around a high. This geostrophic wind causes Atlantic depressions to track northwards across the British Isles. As the depressions approach, their own circulation is augmented by the geostrophic wind, so strong westerly winds dominate.

In summer, the continent heats up, producing a low pressure region over Europe. The geostrophic air flow over the British Isles becomes northerly; approaching depressions track further south and are weakened by the counter current of the European cyclone and strong winds are less frequent.

ALAN CALVERD
BISHOP'S STORTFORD, HERTFORDSHIRE

THE PHYSICAL WORLD

AIR SPACE

Q *We have tried the experiment taught by science teachers, in which a candle standing in water is covered by an upturned glass. The candle goes out and the water level rises in the glass.*

We are taught that the rising water level is caused by oxygen being consumed by the burning candle. However, if we have four candles burning under the glass instead of one, the water level rises much more. Why?
EMMA, REBECCA AND ANDREW FIST
NORWOOD, TASMANIA

A Emma's, Rebecca's and Andrew's questioning of the seemingly well-understood candle experiment demonstrates how young and inquisitive minds are able to demolish false explanations propagated through school physics over the decades.

The consumption of oxygen may well contribute to the rising water level to a certain extent because a given mole volume of oxygen will burn the wax's carbon into roughly the same mole volume of carbon dioxide and the hydrogen into two mole volumes of water vapour respectively.

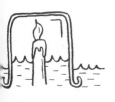

The former will partly dissolve into water, the latter will almost completely condense into liquid water. This will certainly lead to a net decrease in gaseous volume.

However, this is a minor consideration: the important influence is the heat created by the burning candle(s). By the time you cover the candle(s) with an upturned glass, an increased number of candles will have increased the air temperature around themselves more than a single candle would.

As soon as the candle(s) go(es) out, the surrounding air contracts as it cools and the ratio of contraction is directly proportional to the initial average temperature of the air volume under the glass. So more candles lead to more heat, a higher temperature and a higher water level upon cooling down.

All this tells us that we should never believe science teachers without asking a few pertinent questions first.

LEOPOLD FALTIN
VIENNA

 Congratulations to the children who experimentally disproved the common textbook misconception about the candle, the upturned jam jar, the dish of water and the alleged removal of all oxygen from the jar.

By observing that four burning candles cause the water to rise significantly higher up the jar, they have shown that the principal cause of this effect is the heat from the candles, causing the air in the jar to expand. They will also have noticed that the expanded air makes a 'glug, glug' sound as it escapes around the rim. There is a short pause after the candles go out, and only then does the water level rise as the remaining air cools and contracts again.

A candle flame goes out after only a small percentage of the available oxygen has been used up. So it is wrong to claim that this experiment demonstrates the proportion of oxygen in the air in some quantitative way.

IAN RUSSELL
INTERACTIVE SCIENCE LIMITED,
HIGH PEAK, DERBYSHIRE

A The effect is partly caused by the thickness of the three extra candles. You get the same effect using single candles of different thickness. The thicker the candle, the higher the water will rise.

The water drawn in is squashed into the space between the candles and the glass. The narrower this space, the higher the water will rise.

PETER MACGREGOR
GREENOCK, STRATHCLYDE

CLEAR BLUE SEA

Q *Are the oceans blue, and if so, why? I've heard that the colour of the sea is determined by reflection from the sky but it appears blue from outer space and from the perspective of deep-sea divers. And fresh water in a swimming pool looks green. What are the absorption mechanisms involved? Is pure water perfectly clear?*

JOHN RODENBURG
CAMBRIDGE

As you might expect, the colour of the sea depends on both the colour of the light falling on it and the colour of the water itself . . . as well as whether you are looking at it from above or below—Ed.

A The oceans appear blue because of the reflection of light from the sky. The sky appears blue because light from the Sun is scattered by small particles on its way through the atmosphere. The amount of scattering depends upon the wavelength of the light and consequently blue light is scattered the most, which is why we see a blue sky. This also explains why the sky appears red at sunset. At sunset the light has more of the atmosphere to pass through before it reaches us, so the blue light is completely scattered away and only red light remains.

This can be demonstrated by adding a few drops of milk to a glass of water and shining a torch through it. The mixture appears blue when viewed at a right angle to the light source, and red if you look straight into the light source.

BENJAMIN WARD
LOUGHBOROUGH, LEICESTERSHIRE

If you want to demonstrate scattering for yourself, you'll find a simple experiment in 'Take sixteen elastic bands' (*New Scientist*, 26 July 1997, p. 44)—Ed.

Blue refracts and scatters more strongly than green, which affects the colour as seen from above the water. However, the colour of lakes and seas also owes something to impurities, to various colloids or compounds in suspension or solution, and to algae, cyanobacteria or diatoms. For some idea of this effect, sit at the window of an aircraft passing over the ocean on a bright day, look at the colours and try to explain them to yourself. Over open ocean you get deep blue, because there is little in the way of certain plant foods. But as you get to areas such as those close to the Cape Verde Islands off the West African coast, where the prevailing winds blow wisps of rusty dust from the islands into the iron-impoverished sea, you find streaks of green where photosynthetic life is taking advantage of the windfall. Closer inshore the shallow green sea bed dominates the colour.

The mechanism is the same as any colour generation by selective absorption, refraction and scattering. Water molecules and many dissolved substances, such as iron or copper salts in domestic water, absorb some red and orange frequencies more strongly than the rest of the spectrum. The residual light therefore looks green to blue. Teams of divers that spend many days under a few tens of metres of seawater become starved of red light. For some time after they surface into what we see as white light, they see their surroundings as unnaturally red.

Jon Richfield
Dennesig, South Africa

A If you have a colorimetric tube and put only plain deionized water inside, by golly, water itself has a light-blue colour. You can also see the light-blue colour if you fill a sparkling white tub with lots of water.

So pure water isn't optically perfectly clear. If it was, then phone companies laying the optical-fibre cables could use tubes filled with water, rather than purified quartz or optical plastic.

Hiro Sawada
Mississauga, Ontario

DEFLATION POLICY

Q *Why do helium balloons deflate so quickly? When my children bring balloons home from parties, the ones that are filled with helium are often small and wizened by the following morning. I realize that some of the size reduction is caused by deflation but something else must be at work because*

standard air-filled balloons stay inflated for much longer.
JOHN STORR
GREAT CORBY, CUMBRIA

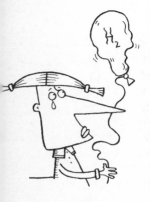

A Helium gas is not only very light, it is monatomic —its particles are all made of a single atom. As a result, helium is made up of the smallest gaseous particles possible. The atoms are only 0.1 nanometre in diameter, and are quite capable of diffusing through metal films. Because it so readily diffuses through small pores, helium is used to help test for leaks in industrial and laboratory vacuum systems. Nitrogen and oxygen molecules have a much larger diameter than helium atoms which means that they are much less capable of diffusing through the balloon walls. It's like the difference between trying to get sand and small pebbles to pass through a sieve— the sand goes through much more easily because it's made from smaller particles.

The second factor which helps to increase losses by diffusion is that balloons are made from viscoelastic materials whose structure is a tangled mass of polymer strands—a bit like a plate of spaghetti. The polymer strands cannot pack closely together, and have channels through which the helium can diffuse, so even at low pressure the helium will diffuse out through the walls. When the balloon is inflated, the polymer stretches, so the balloon walls become thinner (the helium has a shorter distance to diffuse out); the molecular structure becomes slightly more open (making diffusion much easier); and the increased pressure provides a driving force for the diffusion. These are the reasons why deflation is very rapid to begin with, but then gradually slows down as the balloon gets smaller.

Commercial helium balloons are made from non-porous inelastic materials and are coated to reduce the losses even further, although even they still lose a significant percentage of helium per day, certainly enough to disappoint children (and grown-ups) the morning after buying a balloon.

GAVIN WHITTAKER
HERIOT, BORDERS

A The helium atom is very small and very light. It is able to diffuse through the thin, stretched rubber of the balloon quite easily, finding its way through atomic-size pores. Air molecules (oxygen and nitrogen mainly) are larger and heavier and diffuse through much more slowly. In addition to the increased pressure inside the balloon which pushes helium out through the sides, there is another factor that increases helium flow outwards.

Because there is almost no helium in the air, far more helium atoms are hitting the inside of the balloon than the outside, and there is a net flow outwards. However, you will notice that the balloon does not completely deflate. This is because some air moves in as, conversely, more air molecules hit the outside than the inside.

This leads to a truly bizarre effect if the balloon is filled with the gas sulphur hexafluoride, which has large, very heavy molecules which hardly diffuse through the rubber at all, and so cannot get out. In this case, as in the helium example, there are more air molecules outside than in, so air diffuses inwards and the balloon slowly increases in size.

HARVEY RUTT
UNIVERSITY OF SOUTHAMPTON

LONG-DISTANCE CONE CALL

Q *As a child in New Zealand, I was told that the Krakatoa eruption in Indonesia in 1883 was heard in Auckland. (At the time, the sound was allegedly interpreted as the Russian Navy having a spot of artillery practice). By what mechanism could the sound of the eruption remain audible after being transmitted for several thousand kilometres?*
MICHAEL BUCKLEY
ZETLAND, NEW SOUTH WALES

A I understand that the explosion accompanying Krakatoa's eruption was caused by a large volume of water superheating and rapidly turning to steam. The initial eruption created a red hot underwater cavern, which the sea water penetrated. The cavern then resealed as the eruption progressed.

The power of such an explosion was brought home to me and many other keen young scientists of the late 1960s by a lecturer from the University of Nottingham, who toured the country giving a demonstration on explosives. Just before the start of his talk, he placed a small sealed length of glass tube filled with water over a Bunsen flame, surrounded this with thick safety glass, then began his lecture. After several minutes, the audience was brought rapidly to attention when the tube exploded with great noise and violence. The tube had heated up until it reached a temperature at which the glass softened. At that point, the water inside vaporized instantly. As they say on TV: 'Don't try this one at home'.
ANDREW DEARING
THE HAGUE, NETHERLANDS

A There are two parts to the stock answer. First, the eruption was very loud to begin with. Second, the sound was channelled round the Southern hemisphere by the presence of an inversion layer of cold air above warmer air next to the ocean. Such effects do occur on occasion, but usually not over such large distances.
JOHN WOODGATE
RAYLEIGH, ESSEX

A I would like to correct a detail in the answer of a previous correspondent. A temperature inversion is when cold air is found below, not above, hot air. This means that the speed of sound is faster at higher altitude, and for this reason sound tends to curve downwards instead of upwards. The reason this is called an inversion is that the temperature usually drops with increasing altitude.
ERIC KVAALEN
QIRYAT BIALIK, ISRAEL

SHAFTED

Q *If you find yourself in a free-falling lift is there any action that you can take to reduce the effect of the collision? Would jumping just before you hit the bottom of the lift shaft help?*
NIGEL OSBORN
AMERSHAM, BUCKINGHAMSHIRE

A First of all, Hollywood clichés notwithstanding, it's almost impossible for a lift to fall down its shaft, thanks to Elisha Otis's 19th-century patent acceleration-sensitive safety brake. The instant a car starts to fall, multiple spring-loaded arms pop up and wedge it in its shaft.

As for improving your survival prospects, probably the best thing you could do is lie face-up with your back on the floor and your hands under your head to minimize the impact, although this would be difficult to do if you're in free fall.

Jumping just before impact would merely delay your own impact by a few milliseconds. Besides, how would you know when to do it? If you jumped a moment too soon, first you'd bang your head on the ceiling, then you'd be slammed to the floor when the lift hit the bottom.

Even if you could time your jump precisely, to do any good you'd have to exert the same force required to jump to the height the lift fell from (for example, if the lift fell 100 metres, only someone capable of jumping 100 metres in the air could save themselves that way). If they could do that they probably wouldn't need the lift.

Keith Walters
Schofields, New South Wales

A If you jumped a moment before hitting the bottom, giving yourself an initial upward speed, relative to the lift, equal to that of its downward velocity, you would head swiftly towards the roof of the lift compartment. There would be problems with jumping as you would probably be weightless; but with handles to let you pull against the floor it ought to be possible.

Fortunately, just before you hit it, the roof would suddenly accelerate very quickly away from you (assuming the lift kept its shape after impact!) until it had the same (relative) upward velocity as you. Also the floor would do the same, but towards you. You could then land lightly from a few inches off the floor and walk out of the lift onto the ground floor, which would be travelling upwards at the same (relative) velocity.

However, there are one or two problems with this. To achieve such a velocity, you would have to be capable of jumping the height the lift dropped from. And even if you could do this, one surmises the acceleration produced in order to jump would be comparable to that experienced on hitting the bottom.

Even so, by similar reasoning, you would suppose that even a small jump would lessen the impact.

ALEX WILSON
TUFFLEY, GLOUCESTERSHIRE

I can see three ways of increasing your survival chances, although only slightly. One has already been mentioned—jumping as vigorously as possible before you land in order to cancel out some of the upthrust. The second is to get any soft objects you have with you, your clothes for example, and place them underneath yourself prior to impact. This would increase the deceleration time of the collision, and slightly reduce the amount of damage done. If you aren't bothered about your legs, I suppose you could try standing up, and have them act as 'crumple zones', although this could be fairly messy. The third is hardly worth mentioning. You could try to spread yourself out as much as possible while holding on, in order to increase the lift's surface area. This should decrease its terminal velocity by some indiscernible amount.

DAVID FOALE
TOLLERTON, NOTTINGHAMSHIRE

STUMPED

Q *My four-year-old child was wondering if two parachutists would be able to play baseball, cricket or even a simple game of catch while falling. What would happen to the ball?*
ANNI SPRING
ST PAUL, MINNESOTA

A It would not be possible to play a ball game in free fall before parachutes are opened because of the different terminal velocities of the objects involved. The terminal velocity of an object is directly proportional to its surface area–to–weight ratio. For, say, a tennis ball, this is vastly different to that of a person, and the ball would appear to shoot upwards in relation to the sky divers because its terminal velocity is lower (low weight to relatively large surface area).

Also, the very act of throwing a ball would affect the sky diver's control (they use hands to keep stable), and they would be liable to spin around or flip over, which could be dangerous.
M. BURNS
CROWTHORNE, BERKSHIRE

A The parachutists will reach a certain steady downward terminal velocity which is slower than free-fall terminal velocity soon after opening their canopies. Because the parachutists are not accelerating, the rules of motion under gravity are just the same as if they were stationary (just as once a lift has started descending, you can't tell that you are moving downwards). Also, the relative differences of the parachutists and ball terminal velocities is much closer than in free fall.

However, because they are descending, their game will be played in an updraught of air of a few metres per second. So any game would be just like a terrestrial game played in a vertical moderate breeze. Rough sums show that with a dense ball like a cricket ball, they will be hard pushed to notice any difference; a beach ball would travel in a lazier arc than usual, but wouldn't be blown away altogether.

B. CRAVEN
BY E-MAIL, NO ADDRESS SUPPLIED

A It all depends upon the size of the parachute attached to the ball.

CHRISTOPHER GARDNER-THORPE
EXETER, DEVON

CARS IN DRAG

Q *1) When racing cars, motorbikes or bicycles 'slipstream'—follow closely behind each other on a racing track—there is undoubtedly a benefit for the trailing competitor of reduced wind resistance and the pull of the vacuum created by the vehicle in front making a hole in the air.*

However, I'd like to know if there is a corresponding cost to the leading vehicle in having a competitor following closely behind.

DAVID GOLDBLUM
WHITEWATER, WISCONSIN

Q *2) My car is small and so slow that even trucks and buses outpace me on the local highway.*

When such a vehicle approaches from behind, I always experience a very distinct drag at a point when the nose of the large vehicle is still perhaps a metre or two behind my car.

It seems that a wave of air in front of the larger vehicle pulls me towards it, instead of pushing me forward, as one might expect. The turbulence around the sides of the large vehicle, on the other hand, does indeed push me sideways and away from the passing vehicle. What's going on? And is the shape of my car important in this worrisome process?

JAN ROGER
SANDBAKKEN, NORWAY

The first effect described is well known to motor-cyclists. When you approach a large, fast-moving vehicle from behind, you notice a number of distinct zones of air pressure around it, each with its own properties.

The most famous is the slipstream, the low-pressure area just behind the vehicle. This forms because the air that the truck pushes out of the way in front of it lags a little when it sweeps round to fill the truck-shaped hole behind the vehicle. If you are sufficiently young or foolhardy it is possible to tuck your motorbike into this very still area, where there is no head wind and the air dis-placed by the truck's passage is actually pushing you forward as it tries to rush from behind into the ever-departing gap. I used to be able to get an extra 15 kilometres per hour out of my first 2-stroke motorcycle. This is especially useful on long upward inclines, if you aren't worried about riding less than a metre behind your 'tow' with absolutely no hope of surviving if it brakes.

REINHARD READING
NUTBORNE, WEST SUSSEX

A constantly changing, partial vacuum is always present just behind any moving object, because air rushing in to fill the gap takes time to get there. This vacuum pulls the object backwards. This is called drag. The faster the object, the greater the vacuum and the more drag.

A bus slowly gaining on you pushes a band of high-pressure air before it and feeds this into the low-pressure area behind your car, decreasing the vacuum temporarily. Slightly less drag means you will increase your speed a little. Shortly afterwards, at a critical point (which is very sudden), the bus experiences the pull of your vacuum and simultaneously increases this dramatically by blocking off the in-rushing air with its large, flat front. Both vehicles are 'sucked' equally towards this extra-low pressure area. However, because your car is much lighter, the net effect is that immediately after speeding up slightly, you will decelerate much more than the bus accelerates—it will feel as though you are being yanked backwards.

Then, as the bus pulls out to overtake, it first breaks the increased vacuum in your wake, pushing you forward. Finally, it subjects your offside rear to the band of high-pressure air spilling around its front near side, pushing you away.

DOROTHY REICH
NEWCHURCH, ISLE OF WIGHT

 Much of the energy loss experienced while moving through fluids happens at discontinuities. This is particularly true at the front and back of vehicles, so close coupling usually helps not only the following vehicle, but also the leading vehicle.

If the vehicles are suitably close, the gap between them can be filled with a pocket of air, making them behave almost as one long vehicle with the front vehicle paying for the energy losses from the front and the rear vehicle paying for the energy losses from the rear, and each sharing the cost of the relatively small energy losses from the middle.

KEITH ANDERSON
KINGSTON, TASMANIA

A cyclist travelling at high speed (about 40 kilometres per hour) on a flat road expends up to 90 per cent of leg power overcoming the aerodynamic drag and is therefore quite sensitive to small variations in wind resistance.

This resistance is determined by the density of the air the cyclist travels through, the square of his velocity and also by his frontal area and general geometric shape. In calculations of drag, these are taken into account by a drag coefficient.

Everything else being equal and within reasonable limits, this coefficient tends to decrease with increasing lengths of the body travelling through the air. For a 2-metre-long cylinder with a frontal area of half a square metre travelling with its axis parallel to the direction of motion, the coefficient will be 0.93. This will be reduced to 0.83 if the cylinder's length is doubled. While a cyclist is hardly a perfect cylinder, body length is effectively doubled if a companion follows close behind.

RADKO OSREDKAR
UNIVERSITY OF LJUBLJANA,
SLOVENIA

With a closed-up line of bicycles travelling at 25 miles (40 kilometres) per hour, all the slipstreaming cyclists will experience an energy saving of between 26 and 27 per cent. None of the slipstreaming bicycles would have any effect on the bicycle ahead of it.

   With a line of four cyclists riding at 25 miles per hour and taking equal turns at leading, the energy each would use would be the same as if cycling independently at 22 miles per hour. This information comes from an article by James M. Hagberg and Steve D. McCole in *Cycling Science*, September 1990.

ALEX ROTHNEY
EAST GRINSTEAD, WEST SUSSEX

By being very close at high speed, racing cars (notably the American NASCAR stock cars that race on huge ovals with banked turns) share the drag, allowing two or more cars in a line to ease up on the throttle while maintaining speed. When accelerating to overtake, the second driver will pull well clear of the front car, making that car experience all the drag alone and causing it to drop back and be more easily overtaken.

ALEX TAYLOR
SANS SOUCI, NEW SOUTH WALES

INSIDE MACHINES

AERIAL
BOMBARDMENT

Q *The television aerial on my chimney has the classic shape—it has a small metal grid at one end from which a long rod protrudes with evenly spaced spines along it. Why are aerials shaped like this?*

DAVID LOMAX
BY E-MAIL, NO ADDRESS SUPPLIED

A The aerial is a Yagi-Uda array, named after two Japanese electrical engineers: Hidetsugu Yagi, who invented it in 1928, and Shintaro Uda, who perfected it for television reception in 1954.

In this sense, array means having many elements, which is exactly what this type of aerial has. The metal grid at one end is a reflector. It focuses intercepted television signals on to the actual aerial in front of it, which is called the driven element, and this part of the array is connected by a cable to your TV.

The principle is much the same as that of a silvered reflector at the back of a car headlight, which focuses the light emitted by an electrical lamp in front of it into a beam to light the road ahead.

In front of the driven element, away from the reflector, are several director elements, which are the spines you describe. They sharpen the focus of the aerial in one direction, much as the lens of the car headlight directs the light from its reflector into a particular pattern.

For your TV aerial, this pattern resembles a cone, which should be pointed directly at your local TV transmitter to receive signals from it. The rod along the axis of the aerial supports the elements mounted transversely on it and keeps them in their correct places, much as the frame around the car headlight holds the reflector and lens together.

The reason why the Yagi-Uda array has proved so successful as a VHF (very high frequency) and UHF (ultra high frequency) television receiving aerial is that, for such a small size, its performance is excellent. A similar performance from a parabolic aerial, of the type used to receive satellite television signals, would require a dish many times larger.

MICHAEL BRADY
OSLO

The surprising thing about the Yagi-Uda array television aerial is that most of it is not connected to the television at all. The only part that is linked to the receiver, via the coaxial cable, is the single pair of spines which comprise the dipole and lie immediately in front of the small metal grid or reflector.

The length of the dipole itself needs to be about half a wavelength in order to receive the television signal, but on its own it would be too weak in most areas. The reflector is a quarter of a wavelength behind the dipole, and adds significantly to the gain—or strength—of the aerial. All the other spines, known as directors, are in front of the dipole and they, too, are a quarter of a wavelength apart and half a wavelength long. Their job is to increase the gain of the aerial—so the weaker the signal, the more are needed.

There is a price to pay for this increased gain, and that is the directional properties of the aerial. A short aerial with few directors in an area close to a transmitter needs little care in aligning it, but a large aerial that is carrying the many directors necessary in areas with a weak signal, needs to be aimed very carefully at the local transmitter.

Terrestrial television transmission is UHF, which has a short wavelength, so the half-wave dipoles are short. FM radio is VHF which has a longer wavelength, so the half-wave dipole needs to be longer. So, even though some FM radio aerials are similar in design, they need to be much larger.

Some television aerials have their spines, known as elements, in a horizontal plane—this means that the local transmitter is using horizontal polarization, while a vertically aligned aerial means that the transmitter is using a vertically polarized signal.

This arrangement helps TV watchers to avoid interference between transmitters which are located close together. FM radio transmission uses mixed polarization (horizontal and vertical) because many of the listeners are mobile, using portable radios or radios in cars, and are unable to maintain regular alignment of the aerial.

PETER BUCK
PHYSICS DEPARTMENT, EXETER COLLEGE

FAN POWER

Q *I use many fans to cool my airless, New York apartment. One of my fans has two blades, one has three blades and two others have five blades.*

What considerations determine how many blades a fan has? How does the number of blades a fan has affect the amount and velocity of the air it moves?

***And, as an aside, are the underwater
propellers that are used for moving boats and
ships subject to the same considerations?***
ARNOLD KLEIN
NEW YORK

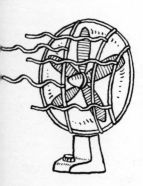

The volume of air moved by a fan is in proportion
to the surface area of the blades in the fan, while
the speed of the wind is determined by the angle
of the blades and their velocity.
BY E-MAIL FROM
BUNDABERG, QUEENSLAND

The number of blades in a fan depends upon
three primary considerations: how much power is
available to turn them, the volume and velocity
of air required, and the restrictions imposed
by noise.

Having more blades is less efficient than having
fewer blades, but a fan with fewer blades needs
to turn faster to move the same volume and will
consequently have a higher tip speed and be
noisier.

Basically, assuming that there are no power
limits, fewer blades on your fan mean more
revolutions per minute, faster moving air, higher
noise levels and maximum efficiency. Many
blades mean fewer revolutions per minute, a
higher volume of air moved, lower noise and
lower efficiency.
JON RADFORD
RAD AVIATION,
OXFORD

The more blades you have in a fan or propeller,
the less efficient it is, because each blade is
to a large extent travelling in the wake of its
predecessor. Indeed, you get the highest
efficiency with a single blade.

In the 1930s and 1940s, when there was a vogue for building and flying model aircraft in competitions, many such hobbyists (myself included) designed and built single-blade propellers, suitably counterbalanced. They even appeared on one or two full-sized aircraft.

The main reason for using multibladed propellers and fans is to reduce their diameter, and hence their tip speed and noise and, in the case of ships, to reduce the turning speed and consequent liability to cavitation, or bubble formation.

GRAHAM SAXBY
WOLVERHAMPTON, WEST MIDLANDS

FLIGHT PATH

Q *Why do rockets take off standing up? Why don't they go along a runway?*
TOBY BIGGINS-GILCHRIST
SYDNEY, NEW SOUTH WALES

A Planes and rockets share a common means of propulsion: both generate thrust by expelling hot gases from a nozzle. The jet/rocket engine pushes the gases backwards, which in turn push the plane/rocket forwards. This is much like jumping off a boat—if you jump one way, the boat will move the other.

However, the key to the answer is that unlike planes, which take off along a runway, rockets do not have wings. A plane needs its wings to fly. The wings have, roughly speaking, a curved top and a flat bottom. If they move through the air, the air has to stream faster around the top than around the bottom. Faster stream velocity means lower pressure, so there is a higher pressure

underneath the wingx than above. As a result
the wing, and with it the plane, gets pushed up.

A plane is rather heavy, so it needs air
streaming past it at very high speed to lift it up.
It achieves this by travelling along a runway
until it is moving fast enough to take off.

A rocket, however, has to travel far from the
Earth where the air is very thin, if it is present at
all, and consequently is not very good for lifting
things. So a rocket cannot use wings to get where
it is going. Instead, it just has a very powerful
engine expelling lots of gas, which does the lifting.
Hence, a rocket does not need to speed along to
get airborne, it just goes up.

Why do we bother to have both planes and
rockets, when just rockets would do? The answer
is that it takes much less fuel to launch a plane
because the air does the lifting.

Seb Jester
Oxford

 Anything that flies in the air must provide itself
with lift. For aeroplanes, birds, insects and so on,
this lift is provided by the air travelling over wings.
Rockets don't have wings, so all their lift must
be provided by the thrust from their engines.

And it's obviously much more efficient to have
the engines pointing straight down so that the
rocket travels straight up, instead of wasting
thrust by travelling horizontally along a runway.

The space shuttle does have stubby little
wings, but they are only useful when it has no
fuel on board and is landing. They don't provide
anywhere near enough lift to get the shuttle off
the ground with its enormous external fuel tank
attached.

Stewart Lloyd
Brigg, North Lincolnshire

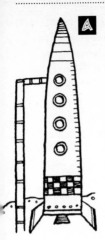

The previous answers to this question have explained why aeroplanes take off horizontally but not why rockets take off vertically. On an airless planet, a rocket could take off in any direction and reach escape velocity using the same amount of fuel. On Earth, however, rockets generally take off vertically to minimize the drag experienced in escaping from the atmosphere.

I recommend reading the early novels of Robert Heinlein for a sound basic course in interplanetary rocketry.

ROBERT ALCOCK
ENERGY RESEARCH GROUP,
UNIVERSITY COLLEGE, DUBLIN

To get into orbit, a rocket must reach a height of about 160 kilometres and a speed of Mach 25. The most efficient method is to go straight up while the vehicle is travelling relatively slowly, and then allow the speed to increase having passed through most of the atmosphere.

A rocket on a runway would accelerate quickly at first, but it would need extra fuel to force its way up through the lower atmosphere at high speed.

ROBIN EMLEY
LIVERPOOL

Dense sea-level air provides great resistance. The only solution is to get above the dense layers of atmosphere as quickly as possible before high speed is reached. The only way to do this is via a vertical take-off.

On the airless Moon, a near-horizontal take-off would have great advantages, especially as some of the driving force could be supplied by a linear induction motor rather than a rocket. This is much more efficient because much less energy goes into accelerating the reaction mass (its fuel, that is). This use of a catapult was first suggested, so far as I am aware, in a science-fiction story by Robert Heinlein, called *The Moon is a Harsh Mistress* (1966).

Of course, on Earth, you could first lift the whole device above the denser layers of atmosphere by using a truly gigantic helium balloon.

ROBERT BAKER
COLCHESTER, ESSEX

IN TUNE

Q *While listening to the radio (on a sound system) tuned to 98.6 FM, I realized that the radio signal could be switched on and off as I moved to different parts of the room.*

It was so sensitive to my position that by sitting on a chair about 6 metres from the radio it was possible to switch the sound on and off simply by leaning forwards and backwards. What caused this phenomenon?
DAVID MARTIN
HAMILTON, NEW ZEALAND

A It sounds as if you are using an aerial which is in the same room as your sound system. In addition to receiving the signal direct from the transmitter, such an aerial will also pick up signals reflected from the walls, ceiling and floor of the room, and from objects in the room, including a human body.

These various signals will interact to form an interference pattern, with the result that the signal strength will vary with the position of any people and the aerial in the room. The wavelength of a signal at 98.6 megahertz is about 3 metres, so a significant phase change can be produced by a person in a room moving less than a metre.

The signal is reaching the aerial directly from the transmitter and also from your body. When these two signals reach the aerial they add together, and the signal the aerial picks up is a result of this addition.

Because the signals are sine waves of the same frequency, the strength of the signal produced by adding them depends on the relative phases of the two signals, the phase difference being caused by the fact that the two signals have travelled different distances from transmitter to aerial because they took two different paths.

If the path lengths differ by an odd multiple of half a wavelength, then the signals arriving at the aerial will be exactly 180 degrees out of phase and the resultant signal, if the two original signals are of exactly equal strength, will be zero.

This shows how you are able to vary the strength of the signal, but how is it that you can switch the sound on and off? Well, most FM receivers include a circuit which mutes the sound output when the signal falls below a certain strength—this is intended to eliminate background hiss when tuning between stations.

What is happening is that, as you move about the room, you are varying the signal above and below the strength at which the muting circuit operates, and hence turning the sound on and off!

Of course, all this illustrates the importance of using a good outside aerial for good-quality FM radio reception. Unfortunately, satisfactory results will rarely be obtained with an aerial in the room.

RICHARD DAVIS
MERLIN COMMUNICATIONS INTERNATIONAL,
LONDON

A The reason for your problem, in which you are not unique, is that FM radio signals, of wavelengths of about 3 metres, are being reflected off the walls, floor and ceiling of the listening room and, of course, off you.

By trial and error, you can place your body in a position to cause this cancellation effect or in a place where you can hear the signal clearly.

However, this phenomenon is not to be confused with the complaint, often heard from radio listeners, that their signal goes off-tune when they move away from the radio. That problem, quite widespread at one time, was caused by a well-known make of radio suffering from cross talk between its AGC (automatic gain control) circuitry and its AFC (automatic frequency control) circuitry.

In this case, moving the body changed the signal level reaching the radio by the means described above, but the radio then altered the tuning position by itself in an attempt to compensate for the problem. Needless to say, this was considered by the owners of the radios concerned to be a fault of FM reception, and thus the responsibility of the broadcaster—in my case the BBC.

CHARLES HOPE
FORMERLY OF BBC ENGINEERING INFORMATION,
WOODSIDE, SURREY

LEAN VIEW

Q *Why do aeroplanes have such small windows, and why are they positioned so low in the fuselage that most people have to*

***bend down in order to see other aeroplanes
on the tarmac?***
TIMOTHY KOULOUMPAS
NEW YORK

As with many things concerning the design of an aircraft, the final arrangement of various parts is based upon a series of compromises. An aircraft designer's life would be so much easier if there were no windows at all, but so far the consensus seems to be that we should have them.

Britain lost the initiative in jet airliner manufacture when the development of the de Havilland Comet in the 1950s suffered a setback through a series of crashes, in part because metal fatigue around the windows led to structural failure.

While windows remain an accepted part of aircraft design, they have since been kept as small as possible. These days they are typically 33 centimetres high. The window has to have three panes: two pressure panes and one interior pane to prevent passengers getting at and damaging the vital ones. The panes are contained in a window unit which is fastened and sealed to the aircraft structure.

The unit is, of course, much heavier and costlier than the thin sheet of aluminium it replaces, and the structure of the aircraft needs to be reinforced to support it. All this extra weight means fewer passengers or less cargo can be carried, so it reduces airlines' potential revenues.

Windows also present a maintenance problem. As well as getting scratched and broken, they are a source of air leaks from the cabin and they also suffer from condensation and icing.

The position of the windows varies depending on the aircraft but generally designers try to place them with their centre line a little below the eye level of seated passengers. On the ground this is perhaps too low, but in flight it gives an oblique view of the ground. Little would be gained by positioning the window higher. Because the seats are placed at the widest part of the circular or oval fuselage, the windows would end up angled upwards some 10 or 15 degrees. The only view the passenger would then have in flight would be of the sky. Also, if the top of the window were above eye level there would be a constant problem from the Sun's glare and dazzle. Passengers would just end up pulling down the blinds, which would negate the benefit of having a window in the first place.

It would be useful to have them deeper, but, as I have already said, the weight penalty makes this impractical.

It also has to be remembered that every civil aircraft flying today was designed at least ten years ago, and some actually started life on the drawing board 40 years ago. During this time people have changed and seat design has changed. When these aircraft were developed, the structural design—including the position of the windows—was fixed and the window line has traditionally been used as a convenient breaking point to bring pieces of fuselage shell together. This position having been determined, and production lines then set up with the correct tooling, it would be enormously costly to change it.

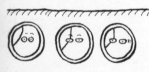

In the meantime, people have been getting bigger. Designers have to use what are known as 'Dreyfuss criteria' to determine seat sizes. These criteria are constantly changing, but a designer will typically make a plane's seats big enough to accommodate 95 per cent of American males. If you are particularly tall, this is going to make the window seem lower for you—and people are generally taller than they used to be.

Finally, the present trend in air travel is away from luxurious, spacious layouts to high-density seating. In these circumstances, where the seat pitch is reduced to accommodate as many passengers as possible, the seat base has to be higher to provide leg room for the person sitting behind. This also makes the relative window position lower still than was originally intended.

TERENCE HOLLINGWORTH
BLAGNAC, FRANCE

 The windows on aircraft are so small to make them safe. The first major jet airliner, the de Havilland Comet, had large, rectangular picture windows through which the passengers had a great all-round view. But after a few years in service, the aircraft started to break up during flight.

To find out why, de Havilland put a new Comet into a tank of water and then pressurized and depressurized it repeatedly to simulate the conditions of flight. After the equivalent of two years' worth of pressurization cycles (which actually only took a few weeks in the water tank), the airframe was found to fail in the top corner of one of the large windows, which caused a catastrophic break-up in flight.

The windows had to be redesigned and small, round windows set low in the fuselage were created. This solved the problem and the position of the windows remains the same today.

MIKE BURNS
WELLINGTON COLLEGE,
CROWTHORNE, BERKSHIRE

MICRO MADNESS

Q *A colleague of mine is in the habit of heating bottled water for his tea in a mug in a microwave oven. When the water is up to temperature he removes the mug.*

On several occasions, the water has started to bubble violently after he has added a tea bag. On one occasion, the boiling started when he was removing the mug. It was so violent that it blew 90 per cent of the water from the mug—which is obviously quite dangerous. What is happening?

MURRAY CHAPMAN
BY E-MAIL, NO ADDRESS SUPPLIED

A A portion of the water in the cup is becoming superheated—the liquid temperature is actually slightly above the boiling point, where it would normally form a gas. In this case, the boiling is hindered by a lack of nucleation sites needed to form the bubbles.

This never occurs when boiling a kettle, for example, because the presence of the rough surface of the element, as well as the convective stirring from rising hot water, are sufficient to produce proper boiling. Turbulence in liquids is known to provide enhanced nucleation in other cases: when you pour a cola drink, for example.

In your colleague's case, the addition of a tea bag (and, in the other case, simple movement) sufficed to allow bubble formation. Even with a large proportion of the water superheated, only a little will convert to steam, because the amount of latent heat required for this phase change is very large. I imagine that by keeping the cup still and microwaving for a long time, one could blow the

entire contents of the cup into the interior of the microwave as soon as you introduced any nucleation sites. It is this sometimes explosive rate of steam production that means you should take great care when using a microwave oven.

RICHARD BARTON
GUILDFORD, SURREY

Superheated liquid can boil explosively if something is added, as in the examples given by your previous correspondents, or if the vessel is moved. I have seen a spectacular explosion of a bottle of liquid which had just been removed from a microwave in a laboratory—glass and hot liquid were thrown across the room. This can be avoided by leaving any liquid that has been heated in a microwave to stand for at least a minute before touching it or opening the door. This allows for slight cooling and for the heat to become more evenly distributed. I recommend that everyone does this when heating liquids in a microwave, even to make a cup of tea.

DIANE WARNE
CAMBRIDGE

ON THE TURN

Why, when driving, does the steering wheel of a car straighten itself if you remove your hands after turning it? It doesn't happen on my friend's Lego Technics car.
CLARE SUDBERY
MANCHESTER

The tendency of the steering wheel to return to the straight-on position is caused by the caster action of the front wheels. This effect is more

clearly seen on a shopping trolley where the vertical swivel axis of each wheel is in front of the wheel-to-ground contact point. If you start pushing the trolley when the wheels are not aligned to the direction of the trolley motion, the wheels are pulled around into alignment by the drag force between ground and wheel.

The full explanation is that as the trolley moves forward, the drag force exerted by the ground on the wheel always opposes any relative motion (or slip) between the wheel and the ground.

Unless the wheels are aligned to the trolley motion, the drag force does not pass through the swivel axis, and therefore it produces a turning moment about that axis which always acts to bring the wheel back into alignment.

In a car, the same effect is achieved by inclining the steering axis and ensuring that the point where the axis intersects the ground is ahead of the tyre-ground contact point. The same is true of a bicycle, as you can see if you hold a broom handle alongside the steering axis of the bike so that the handle touches the ground. You should see that this point is just in front of the tyre-ground contact point.

You can demonstrate the caster action on a bicycle for yourself by pushing the bike backwards and forwards by the saddle while the handle bars are left free. When going forwards, the bike is easy to push in a relatively straight line.

However, going backwards is almost impossible because the front wheel tries to turn round through 180 degrees just like a shopping trolley wheel would. You will also find when reversing a car that the steering wheel loses its tendency to centre itself . . .

BILL LAUGHTON
SOUTHAMPTON

PLASTIC SOLUTION

Q *Checkout operators the world over vigorously rub any malfunctioning credit and debit cards on the nearest available article of clothing. Does this actually serve any useful purpose?*
PHILLIP CLEAVER
NEVRAUMONT, BELGIUM

A From my experience, a credit or debit card will fail to 'swipe' correctly for one of three reasons.

First, something has permanently interfered with the magnetic strip on the card, so that the computer cannot read it. The cashier will have to type in the number manually, and a new card will probably need to be issued. Secondly, the machine is faulty and is unable to read the card.

However, the third reason the card cannot be read is the most common cause. Dust or dirt of some sort collect on the magnetic strip. This obscures the information from the electronic reader. A quick wipe on your sleeve is all that is required to resolve this and, in the vast majority of cases, the card will swipe successfully at the second attempt.

There is no great mystery and no big science behind this practice, at least that I am aware of. If you keep your cards in the card compartment of a purse or wallet, they should remain reasonably clean, and swipe easily on the first attempt. This should also eliminate the first problem, because they will be protected from anything that is likely to irreversibly damage the strip.
CHARLOTTE DADSWELL
PETWORTH, WEST SUSSEX

A There is one drawback to rubbing the magnetic strip and it was something I often experienced as a supermarket supervisor. Rubbing the card can sometimes make it more difficult to read because it becomes charged with static electricity that can interfere with the electronic reader.

The instinct to rub the card in an attempt to remove any dust that may have stuck to it may work in the short term but the extra static charge the rubbing has generated will ensure that even more dust will cling to the card later on.

Cissy Azar
Sydney, New South Wales

Split-Second Timing

Q *How do race organizers time cyclists in the Tour de France? If they do it by video, how do they get separate times from the hundreds of riders who cross the finishing line at once? How did they do it in the old days?*
Richard Kelly
Wellington, New Zealand

A In the Tour de France (and other cycle races), riders who cross the line in a group will all receive the same time as long as there is no gap greater than one second between riders. However, photo-finish equipment is still required to give each rider an accurate finishing position, as time bonuses and points may be awarded to the first few across the line.

To avert the risk of accidents caused by riders all trying to get to the front in the last few hundred metres of a stage, the rules state that if there is a crash in the last kilometre, the fallen riders receive the same finishing time as the main group.

The small time differences between riders in group finishes are not usually considered important in races with different stages as long as the Tour de France, because the time differences between riders on the harder mountain and time trial stages are measured in minutes rather than seconds.

Ian Slack
Wellingborough, Northamptonshire

There has to be a bike length between riders before they get separate times, but they do receive different placings.

Above 15 kilometres per hour the main problem to cycling is air resistance. Riders in a bunch work with each other by slipstreaming behind the leader and changing places, sometimes for many miles, so it would be unfair to give the rider who happens to be at the front at the finish a better time than the rider at the back.

Towards the end of the stage riders will make attacks and try to escape the bunch because once they get away they are very hard to catch again.

Aedan McGhie
By E-Mail, No Address Supplied

The end of a bunch of riders is defined as occurring when there is more than one bicycle's length between consecutive finishers. This is done to prevent all 180 or so riders sprinting to the finish, thus reducing the risk of accidents. This has the desired effect as the non-sprinters in the bunch are quite happy to cruise over the line, in the knowledge that no time is lost.

Andrew Brown
Durham

In the inaugural mountain bike tour of Britain this year, timings were taken by electronically tagging each bike (mountain bike racers are not allowed to change cycles during the race, unlike road racers) using a system similar to the security tags in shops to record the passage of each bike over a line on the ground.

Standard photography techniques are used to determine the final order of cyclists for the purposes of awarding prizes.

STEVE WOODING
SHARNBROOK, BEDFORDSHIRE

THEM!

Q *I have been amazed to see ants emerge seemingly unharmed after being zapped in the microwave, usually after hitching a ride on my coffee cup. They seem to run around quite happily while the microwave is in operation. How can they survive this onslaught?*
JUDITH KELLY
DARWIN, NORTHERN TERRITORY

A The answer is quite simple. In a conventional microwave, the waves are spaced a certain distance apart, because that is all that is needed to cook the food properly. The ants are so minute that they can dodge the rays and so survive the ordeal.
LI YAN
NORWICH, NORFOLK

A The phenomenon that the ants take advantage of is that microwaves form standing waves within the oven cavity.

So in some places in the oven space, the energy density is very high, whereas in others it is very low. This is why most ovens have turntables to ensure that cooking food is heated evenly throughout.

This standing wave pattern can be observed by putting a static tray of marshmallows in the microwave, and heating for a while. The result will be a pattern of cooked and uncooked marshmallows. The standing wave pattern, however, varies according to the properties and position of any material within the oven, such as a cup of water.

The ant will experience this pattern as hot or cold regions within the oven and can thus locate the low-energy volumes. If the ant is in a high-field region, its high surface area–to–volume ratio allows it to cool down more quickly than a larger object while it searches for a cold spot.

It is a common myth that microwaves are too big to heat small objects. The fallacy of this has been demonstrated by chemists such as myself who employ microwave heating in their work. Certain types of catalyst consist of microwave-absorbing particles—typically of submicron size—dispersed throughout an inert support material. There is convincing evidence that microwaves are capable of heating only the tiny catalyst particles.

A. G. WHITTAKER
HERIOT, BORDERS

 There is very little microwave energy near the metallic floor or walls of the oven. The electromagnetic fields of microwaves are 'shorted' by the conducting metal, just as the amplitudes of waves in a skipping rope, swung by a child at one end but tied to a post at the other, are reduced to nothing at the post.

An ant crawling on the rope could ride out the motion near the post, but might be thrown off nearer the middle.

For a quick demonstration of this, place two pats of butter in a microwave in two polystyrene coffee cup bottoms, one resting on the floor, the other raised on an inverted glass tumbler. Be sure to place a cup of water in the microwave as well. On heating, the raised butter will melt long before the butter on the floor.

CHARLES SAWYER
CAMPTONVILLE, CALIFORNIA

UNDER AND OUT

Q **When travelling by car I listen to AM radio broadcasts and it is annoying that the radio signal almost disappears when I pass under any shape of bridge and through tunnels. Why doesn't this happen with FM stations?**
TIBOR WEITZEN
SYDNEY, NEW SOUTH WALES

A The question of radio waves penetrating under road bridges is a special case of the problem of radio propagation down pipes and tunnels. A road bridge is like a very short tunnel.

We can consider three cases; empty tunnels with electrically conducting walls, empty tunnels with walls that are not electrically conducting, and tunnels containing electrical conductors such as pipes or cables running along their length.

In principle, the presence of conductors in a tunnel permits radio waves of all wavelengths to propagate, sometimes a good distance but often perhaps only a few tens of metres, depending on the electrical characteristics of the walls and conductors. Generally, the lower the frequency the greater the distance of propagation.

Tunnels with electrically conducting walls and no separate internal conductors are like conventional waveguides used in microwave radio communication. These will carry only those signals whose wavelength is shorter than twice the dimension of the tunnel cross-section. Microwave signals (perhaps of a wavelength of 10 centimetres) can sometimes propagate for hundred of metres down such tunnels, but they are blocked by corners and bends or by obstructions such as vehicles.

Signals with a wavelength too large to meet the size criterion mentioned above will only penetrate a fraction of a wavelength down the tunnel. In these cases medium-waves might go a little farther than VHF, but not very far.

Normal tunnels with, say, concrete walls but without internal conductors also act as waveguides—but the power of the propagating waves is quickly absorbed by the walls, so the waves don't travel far.

In general, the shorter the wavelength of the radio wave the farther it will penetrate the tunnel before colliding with a wall and being absorbed.

In order to get good radio communication in a tunnel, it is usual to feed VHF signals into a special type of low-loss coaxial cable running down the tunnel's length. The cable is designed to allow a controlled amount of leakage to take place (like a hosepipe with holes in it) so that radio waves are leaked into the tunnel at regular distances. This way each part of the tunnel has its own radio wave supply. This is done in the Dartford Tunnel, where good BBC signals can be heard. Mines and underground railway systems often use the same technique.

QUINTIN DAVIS
LEATHERHEAD, SURREY

WAVY TRAIN

Q *When the paths of spacecraft are being traced on Mission Control's map of the world, why do they show up as wavy lines?*
COLIN MICHAELSON
LONDON

A The track of a satellite's orbit over the Earth's surface is a great circle so it traces a straight line on the surface of a globe. However, the map at Mission Control is what is known as a Mercator projection, in which a spherical surface is distorted into a flat rectangle and the size of polar regions is exaggerated.

On this projection, a polar orbit would appear as a vertical straight line, while an equatorial one would appear as a horizontal line, but any other great circle orbit would look like a sine wave.
PETE FOWLER
SOUTHLAND, NEW ZEALAND

A The wavy lines referred to in this question are the loci of points on the surface of the Earth at which the satellite appears directly overhead at different times. The exact shape of the curve depends on such things as the ellipticity of the orbit, its period and the inclination of its plane to that of the equator.

However, if we simplify the problem by assuming that the satellite has a circular orbit, and a period which is shorter than a day, then the locus approximates to a great circle, and the only variable will be its inclination to the equator.

The representation of a great circle on a flat map (such as a Mercator projection, where the north–south distance from the equator is proportional to the sine of the latitude) is a complicated function of longitude (the east to west distance along the map), but visually is almost indistinguishable from a sine curve.

You can convince yourself of this by tying a piece of string around a model Earth globe on a great circle and then rotating the globe. Imagine that the satellite is following the path of the string at a constant rate. Starting at the point where the string crosses the equator, you will find the path climbing steadily towards one of the poles, but turning round and coming back again before it reaches it. This humped path will then be repeated in the opposite hemisphere. This path is a smooth, sinusoidal(ish), upwards-and-downwards oscillation.

In the case of a real satellite or spacecraft, the Earth rotates to the east under the object as it follows this path, and so the snaky curves are displaced westwards a little on each orbit, and do not join up to form a single curve. Also, the ellipticity of typical orbits mean that the progress around the string is not at a constant rate, so the curve gets squeezed and expanded horizontally (east to west) in different parts as the speed changes relative to the steadily moving world underneath.

ANDREW SMITH
BRISTOL, AVON

 The tracks are wavy (or more correctly sinusoidal) because the orbits are oblique to the equator. You can demonstrate this effect with the cardboard tube from a roll of kitchen paper (which represents the Cartesian projection of the Earth's surface displayed on the wall map).

Hold the tube firmly and, using a sharp knife, cut slowly across the width of the tube at an oblique angle (that is, not at right angles to the tube). Ensure that you do not crush the tube flat while cutting. Then take one of the pieces and make a cut along the length of the tube, parallel to its centre line.

Open the card out and you will see that the cut edge forms a sinusoidal curve.

TONY COYLE
PHOENIXVILLE, PHILADELPHIA

🅐 Another example of this is the type of 'world clock' which shows not only the current times in every major time zone, but also (and more importantly) the terminator which separates night and day. This shadow line also appears as a wavy line on the flat projected map.

BRIAN GARCIA
SAN DIEGO, CALIFORNIA

WHICH WAY IS UP?

🅠 *My whole class, including my mathematics teacher, is baffled. We cannot work out how an aircraft can manage to fly upside down without crashing into the ground. We understand that the wings are designed to provide uplift when the plane is flying horizontally. However, when the plane flies on its back as some smaller jets often do, surely the uplift is working in reverse and forcing the plane back down towards the ground. Yet most types of small aircraft seem able to maintain the upside down position for long periods of flight. How do they do this?*

NIK YUSOKK
LONDON

Although the aerofoil shape of an aircraft's wing produces some of the lift in normal flight, the more important factor is the angle of attack—the angle at which the air strikes the wing.

The wings of an aircraft are normally inclined to about 4 degrees to the horizontal when compared to the main body of the aircraft. This is known as the chord angle of the wing.

 So even when the fuselage is level, the angle of attack into the oncoming wind is 4 degrees. This produces lift in the same way that your hand experiences an upward force when you hold it at about 45 degrees to the horizontal in a fast-moving stream of air. Your hand does not have an aerofoil shape but the lift that you feel is caused by the angle of attack of your palm to the oncoming wind.

It is this principle that allows an aircraft to fly upside down. The nose is pointed further upwards than in standard flight because of the need to offset the chord angle of the wing. But if the angle of attack is positive compared to the relative airflow over the wing, then an upward force will still be produced. It is this lifting force which overcomes the force produced by the shape of the wing, and holds the aircraft in the air.

The bigger problem that pilots should be concerned about when flying their aircraft upside down is the risk of the engine stopping, because both the oil and fuel systems in most ordinary light aircraft are fed only by gravity. Flying your aircraft upside down can easily cut off the fuel supply because the valve that is feeding fuel to the engine suddenly finds itself at the top of the tank.

MARK MOBLEY
BRISTOL, AVON

HOUSEHOLD SCIENCE

ACID TEST

Q *How does putting a few drops of vinegar into water that is being used for boiling eggs prevent the eggs cracking?*

SANDY FUNG
CAMBRIDGE

A Adding salt and vinegar to the water in which eggs are boiled does not prevent eggshells from cracking. Shells crack either because they are dropped into the saucepan too roughly, or because the air in the air sac at the broad end of the egg expands rapidly because of the sudden heat increase.

The first of these causes can be avoided by lowering the egg gently into the water on a spoon, the second by piercing the shell at the broad end with a needle before cooking the egg, thus allowing the expanding air to escape without the combined danger of cracking the shell. I suppose cracks could also occur when eggs are boiled very rapidly and they constantly collide with the saucepan base in the turbulent water, but eggs should be simmered anyway to avoid tough, rubbery whites.

The purpose of adding salt and vinegar to the water is to make the egg white coagulates more quickly when the shell does crack, so that less escapes into the water.

RENE THOMSON
ASHFIELD, NEW SOUTH WALES

We don't believe the explanations below answer the original question. But they do describe a great party trick and how it works. If you put only a few drops of vinegar into the water in which you boil an egg, the solution will be too dilute and the

egg's exposure to it too short to affect the shell significantly—Ed.

 A trick I learnt some years ago demonstrates how flexible an eggshell can become when soaked in vinegar overnight. Presumably, it is this reduction in brittleness that prevents the shell cracking when vinegar is added to the water in which an egg is boiled.

First, soak a fresh egg overnight in vinegar, ensuring that it is fully immersed in the liquid. Then choose a bottle with a neck that is narrower than the diameter of the egg—about two-thirds of the diameter of the egg seems to work best.

 You need to be extremely careful with the next step, which involves dropping a piece of burning paper into the bottle and then placing the damp egg on the top of the bottle. The experiment works best if the pointed side is facing down into the bottle opening. Children who wish to do this experiment should be supervised by an adult.

As the air cools, the reduction in air pressure will draw the egg into the bottle. If you then rinse out the bottle and egg with fresh water, the egg shell will be restored to its original shape and hardness. You now possess a conversation piece that will baffle anybody who does not know about this trick. How did a solid egg get into the bottle?

IAN DOCHERTY
BRACKNELL, BERKSHIRE

 Soak an egg in pure vinegar for 24 hours and it will become rubbery and flexible as a water-filled balloon. Try it, it really is quite amazing!

The shell of an egg has the same composition as bone. It has a mineral component (hydroxyapatite) and an organic component (the protein collagen). The mineral component makes the shell hard but fragile, while the flexible collagen acts like a glue that maintains the integrity of the whole structure. Acids such as vinegar dissolve the calcium-based mineral component, which is alkaline, leaving the flexible collagen intact. Hence you get a rubbery eggshell.

PEDRO GONZALES
LONDON

In response to the above correspondent who previously answered this question, we would like to point out that the main mineral component of eggshells is actually calcium carbonate, rather than the calcium phosphate (hydroxyapatite) that is found in teeth and bones. Calcium carbonate is also found in various sea shells and snail shells.

ANA VILLACAMPA AND JULYAN CARTWRIGHT
GRANADA, SPAIN

CANNED EAT

Fresh grapefruit are almost impossible to peel by hand without splitting or tearing the fruit. So how are the perfect skinless segments found in cans produced?

F. G. GRISLEY
BARRY, GLAMORGAN

My brother worked for a while in a citrus plant in Florida. Grapefruit are put in very hot water for a short time, after which the peel can be removed easily with a short-bladed knife. Often it can be removed almost in one piece.

BY E-MAIL, NO ADDRESS SUPPLIED

It's a long time since I worked in a grapefruit canning factory but at the time I believe the whole

fruits were blanched in a hot caustic soda solution.
This made them much easier to peel by hand.
PATRICIA BAYLEY
HUDDERSFIELD, WEST YORKSHIRE

A The thick white peel and outer membranes are
removed by one of two methods. In the first, the
grapefruit are heated for a few minutes to almost
100 °C to make their peel 'puff', after which one
piece of peel is sliced off to start the process. In
the second, the grapefruit is immersed for a few
seconds in hot caustic soda, which is then washed
off with water sprays. The peel can then easily be
removed by hand.
C. RENARD
RENNES, FRANCE

COLOUR CODING

Q *Why is house dust always grey?*
RICHARD COOPER
NORWICH, NORFOLK

A Here in the dry, dusty Middle East, our house
dust is closer in colour to the pale sand outside.
CHANA LAJCHER
JERUSALEM COLLEGE OF TECHNOLOGY,
ISRAEL

A Our dust is red, because we live in north Texas,
close to the Red River. Our land is composed
mainly of red clay.
S. TINNEY
TEXAS

A Under normal conditions, house dust consists
mostly of sloughed-off skin cells. Amazingly, most
of the dust on top of the wardrobe is dried-out
human skin.

Dry skin is a translucent grey colour, and consequently, so is the dust. There's no other colour in it because the blood vessels are much deeper down and they are not lost as the skin grows.

Other forms of dust, such as windblown soil, can be different colours, and the colour gives clues to their origin.

John Morton
School of Applied Sciences,
University of Glamorgan,
Pontypridd

And if we have any readers that question whether all that house dust is really dead skin, here's a proposal for an experimental test—Ed.

Grey household dust is largely human skin. Although humans come in a selection of colours, the pigments are found below the layer that generates our outer layer of dead skin. To test this, perhaps a reader would like to bath in woad and report if the dust turns blue.

Roger Wilkins
Felixstowe, Suffolk

DOUBLE TROUBLE

I recently purchased a box of eggs, each of which was guaranteed to have two yolks. And the claim was correct. How does the supplier ensure that each egg has two yolks?

John Crocker
Solihull, West Midlands

These special eggs are a natural phenomenon over which we have no control. Double yolk eggs are larger than those laid by the majority of the flock and are set aside to be tested individually. Demand for double yolkers far outstrips supply and we need to be very sure that they do in fact contain two yolks. Each egg is therefore checked by holding it against a bright light. During this process (still known as candling from the days when a candle provided the source of light) the number of yolks will be clearly visible as shadows.

GRAHAM MUIR
STONEGATE FARMERS LIMITED,
HAILSHAM, SUSSEX

Try it at home—you'll be surprised how much of the inside of an egg you can see—Ed.

EGGSTRAORDINARY

Q *Why does a very fresh egg take appreciably longer to cook than one that is a day or more old?*
RICHARD MOORE
ASHURST WOOD, WEST SUSSEX

A Clear egg albumen coagulates irreversibly at about 58 °C and becomes white. When cooked in hot water, the heating is achieved by direct heat conduction from the outside, through shell and egg white, and by convection currents in the water of the egg white which contains the dissolved albumen and surrounds the yolk.

A fresh egg has a much larger volume of this clear albumen solution trapped between the mucin fibrils that make up a gelatinous lattice. This lattice forms three concentric dense layers of egg white that surround the yolk. A fresh egg has a deep and strong lattice in its middle dense layer that causes the egg to sit up more in the pan when it is broken for frying. After some days the lattice degenerates and the more watery egg can be seen to flatten out in the frying pan. Convection is much easier in such an egg and the heating of the albumen through to the yolk is accelerated.

STEPHEN TOMKINS
CAMBRIDGE

FRYING PROBLEM

Q *When I view the surface of cooking oil in a pan by reflected light, a pattern of honeycomb-like shapes appears as the pan is heated by a gas flame. The unit size of the pattern is smallest where the oil is thinnest. Why is this?*

REX WATSON
BROADSTONE, DORSET

A The honeycomb cells observed in heated cooking oil are known as Rayleigh-Bénard convection cells. If the temperature difference between the bottom and the top of the oil is low, the heat is dissipated through ordinary thermal transport (collision of individual molecules) and no macroscopic motion can be observed. At greater temperature differences, convection (a collective phenomenon involving many molecules) is a more effective means to transport the thermal energy. The heated cooking oil on the bottom is less dense and wants to rise. The top of the oil cools down by contact with the air and sinks again. This motion becomes circular and creates rolls of liquid, which self-organize into a honeycomb pattern which can be easily observed.

Quite a bit of research has been carried out on this phenomenon, which anyone can create in the kitchen, and we now have an explanation as to why the pattern of the convection cells is honeycomb. The form of the convection rolls depends on the shape of the container in which the liquid is heated. Hexagonal patterns seem to develop easily in round pans. Other containers may lead to long, rectangular rolls, with a square cross section.

As the liquid moves in a circular fashion (up, across, down and back across), the unit size of the pattern depends linearly on the thickness of the liquid. It is interesting that many parameters such as the unit size of the convection cell are determined, whereas the direction of the circular motion is undetermined at the onset of convection. Once a rotation direction (clockwise or anticlockwise) is established, it remains stable.

BERND EGGEN
UNIVERSITY OF EXETER

A Twenty seconds or so after heat is first applied, the really interesting phase of convection begins suddenly. When the temperature gradient within the oil layer has built up to a certain critical value, each of the many scattered convection currents present in the oil finds that it conserves energy better if it shares its region of descending flow with the down-flow regions of its immediate neighbours. This stops any contraflow problems. This cooperative repositioning of the centres of convection forms a regular pattern of closely packed convection cells. Their honeycomb-like appearance occurs to allow each cell to have the maximum area consistent with sharing its cell walls with its neighbours.

Because of this cell cooperation, convection proceeds vigorously and the rising hot oil can be seen to form a small fountain at the centre of each cell. The force that maintains this pattern, in the face of mechanical and thermal disturbance, is the flow of heat energy up through the oil layer. In the same way, a biological system needs energy throughout—in this case food—to maintain its integrity.

A substantial increase in the temperature gradient leads to the break-up of the cell pattern, which may pass through several phases of greater complexity before degenerating into chaos.

ROGER KERSEY
NUTLEY, EAST SUSSEX

It can be shown analytically that the most efficient flow pattern in a large expanse of fluid transferring heat from bottom to top is hexagonal, with cells about the same width as the depth of the fluid. The hot fluid moves up the centre, cools at the surface and then drops down the perimeter of the hexagon. Similar patterns can be seen on all scales from millimetre-sized experiments to patterns on the surface of the Sun.

GARY ODDIE
CRANFIELD, BEDFORDSHIRE

The readers above have already provided answers to this question. However: as our correspondent below points out, the previous explanations using the Rayleigh convection model were not wholly correct, for the Rayleigh model only applies if the frying liquid is of sufficient depth—Ed.

The behaviour of hot oil in a pan is a classic example of Bénard convection, the unstable motion of fluid on a heated flat plate which takes the form of regular hexagonal cells of circulating fluid. It is well known that Lord Rayleigh developed a theory to explain this instability. What is not so well known is that his model was wrong.

Rayleigh considered a horizontal layer of liquid with flat surfaces heated from below, and assumed that the instability took the form of parallel, contra-rotating rolls driven by buoyancy forces due to variations in the fluid density. Then, by heuristic arguments, he deduced a size for hexagonal cells close—fortuitously—to that observed by Bénard. He also predicted the minimum temperature gradient across the layer for the onset of this motion but this was about 100 times greater than the gradient needed to initiate the cellular flow in Bénard's experiments.

Other researchers extended Rayleigh's analysis in various ways. When the flat upper surface condition was later relaxed it could be seen that the surface is elevated above rising fluid between adjacent rolls while it is depressed above descending fluid. This is precisely the opposite of what Bénard observed. When Bénard's experiment was repeated it was found that the cells could also be produced when the heating plate was cooled, whereas according to Rayleigh's ideas the fluid should remain at rest. The instability has also been observed for a layer of liquid beneath a plate being heated from above and in space, where gravity and hence buoyancy forces are zero.

In the late 1950s a new model for Bénard convection was developed in which variations of surface tension caused by temperature variations on the surface of the liquid drove the motion. This model also predicted a depressed surface above rising fluid. In reality both Bénard and Rayleigh effects must be present. Conditions determine which predominates. Buoyancy forces drive the motion when there is no free surface or the liquid layer is thicker than about 10 millimetres; otherwise surface tension governs the flow.

Whichever driving force dominates, it must be sufficient to overcome the effects of viscous drag (which tends to inhibit motion) and diffusion of heat within the fluid (which tends to reduce the temperature gradients) before it can initiate the unstable flow. For buoyancy-driven flows the onset of instability is governed by the Rayleigh number: buoyancy forces/(viscous drag rate of heat transfer); while for flows driven by surface tension the corresponding variable is the Marangoni number in which surface tension forces replace buoyancy forces.

For thin layers the unstable flow takes the form of a regular array of hexagonal cells regardless of the shape of the container. For thicker layers the basic unstable flow is a series of rolls parallel to the container's sides with the direction of flow adjacent to its rim and determined by its temperature relative to its base. These rolls degenerate into polygonal (but not necessarily hexagonal) cells when the temperature gradient is increased.

RICHARD HOLROYD
CAMBRIDGE

IRON STAINS

Q *During my long hours ironing, I have noticed that when I use a hot iron some colours, mainly reds and blues, change shades. Why does this happen?*
ELENA LAMBEA
LONDON

A When a garment, especially an older and more worn one, has been freshly washed, the surface fibres fluff up and become randomly aligned. These fibres can collect and reflect light in a number of directions, making the fabric appear lighter.

However, when they are ironed, these fibres are flattened against the fabric, all in roughly the same direction. Therefore, less light is scattered from the fabric surface and it takes on a darker shade. This is most noticeable on dark colours like green, red and blue.

ANDREW KNIGHT
MANCHESTER

IRONED OUT

Q *What are the physical and/or chemical processes that take place when the creases in clothes are removed by ironing? Is the process independent of the type of material ironed?*
BRIAN PORTER
TRING, HERTFORDSHIRE

A Fabrics contain fibres, each of which is composed of many long-chain molecules lying alongside one another and loosely bonded together. If these bonds are undone and remade elsewhere the molecules (and the fibres they make up) may be forced to run straight and true.

Ironing a cotton shirt provides an example of the process. Cotton is made up of cellulose molecules, polysaccharides, composed of a long chain of glucose-like subcomponents. Hydroxyl groups stick out from the sides of the cellulose molecules and attach to those of neighbouring cellulose molecules by hydrogen bonds.

These bonds can be broken with enough heat and a little water, which causes swelling. Remember how hard it is to get the wrinkles out of a cotton shirt that is very dry, or with an iron that is not hot enough. In one case the fabric does not contain enough water to swell the fibres and easily break the hydrogen bonds, and in the other, the temperature is not high enough. The bonds re-form as the iron is removed, leaving the shirt with the shape pressed into it—nice and flat.

In fabrics other than cotton, the effect is similar, even though the bonding between adjacent long-chain molecules may be different from the hydrogen bonds in cotton. For example, wool also contains covalent and electrovalent cross-links which allow the fabric to be permanently pleated by breaking and re-forming the bonds with selected chemicals.

Polyamide (nylon), polyesters, acetates and triacetates are thermosensitive and can be heat set.

Michael Parkinson
De Montfort University,
Leicester

MELTDOWN

Q *The chocolate chips in a biscuit do not appear to melt when cooked in the oven at 150 °C, but melt when left in the Sun. Why?*
The Lumsden Family
High Wycombe, Buckinghamshire

A The chocolate chips do melt when the biscuit is cooked. The rest of the biscuit mixture is thick enough to hold the bits of molten chocolate in their original shape inside the hot biscuit, giving the impression that nothing has changed when the biscuit cools. Try eating one while it is still warm. You will soon see that the chocolate has melted.

NIGEL GOODWIN
HEREFORD

ON YOUR MARKS

Q *Does a bleaching agent really get rid of stains or does it just make them invisible?*
CORINA LEE
CHICHESTER, WEST SUSSEX

A Practically all bleaching products today contain oxidizing agents capable of rapidly disrupting the delicate chromophoric balance of stains, making them invisible. Then, given sufficient time and temperature, they fragment the whole stain structure into water-soluble pieces, so guarding against any potentially embarrassing return of the offending colour.

Sodium hypochlorite (chlorine bleach) is the most powerful oxidizing bleach familiar in the home and is capable of stain removal at ambient temperature, but it is not used in British laundries, being judged too damaging to fabrics, dyes, enzymes, perfumes and so on. Laundry bleaches in Britain are based on hydrogen peroxide which is much less reactive and hence more discriminating. It performs superbly in a long boil wash but, unaided, is insufficiently reactive to cope with today's short 40 °C washes.

Detergent companies have created ways of activating peroxide to improve performance while reducing its damaging effects. The latest move is to convert the peroxide into peroxyacids during the laundry operation but the favoured approach in the future might be catalytic activation of the peroxide by transition metal ions. However this infant technology is suffering some teething problems.

FRED HARDY
NEWCASTLE-UPON-TYNE

STALE TALE

Q *Why does a biscuit that is left in the open overnight become soft by the morning but a baguette left out for the same length of time become so hard that one could knock someone out with it?*

LORNA HALL
BULLION, FRANCE

A Biscuits contain much more sugar and salt than baguettes. The finely divided sugar and salt are hygroscopic and soak up moisture from the atmosphere—the osmotic pressure in a sweet biscuit is quite high. The dense texture of a biscuit helps maintain the moisture by capillary effects.

The baguette, on the other hand, contains little salt or sugar, and has a very open structure. The flour doesn't care if there's moisture around it or not. So, because of their different make-up, one attracts water, the other doesn't. Try a series of different biscuits, varying from very sweet, dense ones to light, fluffy sponge biscuits. The 'overnight sogginess index' increases with density and sugar/salt content. I find that if I put both traditional Italian biscotti (not very sweet and fairly open-textured) and dense, sweet ginger biscuits in a sealed container, the biscotti go rock hard and the ginger biscuits end up very soft.

CHRIS VERNON
KWINANA, WESTERN AUSTRALIA

A   A baguette dries out while a white sugar biscuit becomes soft because of the hygroscopicity of the white sugar in the biscuit. I researched this last year when entering a competition at the age of 13. We were asked to produce a project about whether cookery was a science.

The water vapour in the air is attracted to the sugar and this makes the biscuit softer. Baguettes however, have no sugar in them and therefore have nothing to attract the water vapour, which evaporates to leave the baguette hard.

When we performed the experiment we used three types of biscuit: one made from caster sugar, another from honey, with the last being the control which had no sweetener. The control lost 2.17 grams of water after being left outside overnight, and the honey lost 2.03 grams, but the caster sugar biscuit gained 1.23 grams. The honey biscuit lost water because the atmosphere had a lower concentration of water than the biscuit.

TOM WINCH
ELY, CAMBRIDGESHIRE

A   Starch consists of about 20 per cent amylose and 80 per cent amylopectin. The key to bread becoming stale is amylose retrogradation. Naturally, loss of moisture is involved or it wouldn't dry out. However, bread can be prevented from losing moisture and still go stale. The linear amylopectin molecules in the starch grains which are separated by moisture in fresh bread, move closer together and become more ordered as the bread becomes stale, making it stiffer.

The process is temperature dependent, with the rate fastest at just above freezing and slow below freezing. Studies show that bread stored at 7 °C (average fridge temperature) becomes stale at the same rate as bread stored at 30 °C. So putting bread in the fridge does not keep it fresher for longer.

ALLIE TAYLOR
LONDON

A The feature referred to in the question has a parallel in legal terms. Here there is a difference between cakes and biscuits for VAT purposes. This is important because cakes are subject to VAT, while bread is not. Now we have a new definition: a biscuit is something which goes soft when left out, whereas a cake goes hard. What the implications are for VAT on baguettes, I wouldn't like to imagine.

RICHARD BUTLIN
LONDON

TOUGH AS OLD . . .

Q *Many years ago, during the First World War, my grandfather was travelling through New Zealand when he came across a farmer wearing one bright shiny boot and one dirt-encrusted boot. When asked why, the farmer said he was testing whether the regularly polished boot would last longer than the one protected by dirt. Does anyone have any suggestions as to which boot would last longer?*

JENNI GYFFYN
MELBOURNE, NEW SOUTH WALES

From my experience as a footballer, boots that are polished regularly last longer and feel softer. Apparently, boots that are not polished become hard, allowing cracks to appear. Polishing also acts as a protective layer between the leather and the outside world.

DIMITRIOS SIGANOS
LONDON

Surely the mud-encrusted boot will be ruined first as the mud will contain bacteria and the like which will identify the leather as a food source. The polished boot will have the protection of the polish, which in most cases is waterproof, and this will prevent the boot from rotting.

IAN NEWTON
BY E-MAIL, NO ADDRESS SUPPLIED

STRINGS ATTACHED

Q **Why does grilled cheese go stringy?**
JOHN MITCHELL
WISHAW, STRATHCLYDE

The uncooked cheese contains long-chain protein molecules more or less curled up in a fatty, watery mess. When you heat cheese, the fats and proteins melt and if you fiddle with the fluid, the chains can get dragged into strings. Grab a bit of the molten cheese and pull, and you get a filament, in much the same way that you can draw and twist cotton wool into yarn.

You can do similar things with polythene from plastic bags by heating or stretching the plastic to curl or stretch the long-chain molecules. When the

molecules are curled up, the plastic is softish and waxy. When they are stretched into fibres, the result is elastic and strong in the direction of the stretch, although it splits easily between the chains lying along the fibre.

JON RICHFIELD
DENNESIG, SOUTH AFRICA

 As the cheese melts, the long-chain protein molecules bind together to form fibres in the liquid mass of melted cheese. I believe that this characteristic can actually be used to measure the protein content of a cheese sample directly. A string of cheese is pulled away from the sample, and the distance to which the fibre will extend away from its attachment point on the main piece of cheese is measured against some reference sample of known protein content.

MIKE PERKIN
BY E-MAIL, NO ADDRESS SUPPLIED

GADGETS
AND
INVENTIONS

COLD STORE

Q *Can I extend the storage life of my torch's batteries by keeping them in the refrigerator?*
TIM GILLIN
RANDWICK, NEW SOUTH WALES

A Primary batteries such as the Leclanché (zinc–carbon) or alkaline–manganese systems can have their storage lives extended significantly by storage at 0 °C. For such cells, the loss of capacity on open circuit is caused mainly by chemical side reactions such as zinc corrosion. Such processes, like chemical reactions in general, are accelerated by a rise in temperature. Water loss is affected similarly.

For a Leclanché cell stored at room temperature the capacity will fall to about 90 per cent after a year, to 70 per cent after two, and to 40 per cent after three. At the much higher temperature of 45 °C, it will have its capacity reduced to 20 per cent after only a year. However, at −20 °C the capacity will have fallen only to some 80 per cent after 10 years. Therefore, it is best to store such batteries in a refrigerator. Keeping them in a deep freeze, however, could damage the batteries if the can and seals have significantly different coefficients of expansion.

Lithium batteries have longer shelf lives because when the shell is filled during manufacture, the lithium anode reacts rapidly with the electrolyte to form a compact passivating layer. This virtually stops further corrosion. Even when stored at room temperature, they can retain up to 90 per cent of capacity after 10 years. Some cells (lithium–sulphur dioxide, for example) retain more than 70 per cent of their full capacity even after a year's storage at 70 °C.

In contrast, aqueous secondary cells have very poor charge retention—especially nickel-cadmium and nickel-metal hydride batteries. At room temperature the available energy of a recharged battery drops to about 50 per cent after six months' storage for new, low-discharge rate, sealed cells and to even lower values for cells which are high-power or which have been cycled many times. Again this loss may be reduced to about half of the room temperature amount by storing the charged cell in a refrigerator. Better still, keep such batteries connected to an appropriate charger when not in use.

COLIN VINCENT
NEWPORT-ON-TAY, FIFE

CRACKED IT

Q *I am unable to crack a Brazil nut to extract the contents in suitably sized edible pieces. The kernel shatters and I find that the only way that I can retrieve the crumbs is on the tip of a wet finger. How do the processors of Brazil nuts actually get them out of their shells?*

J. R. CATLIN
SEASCALE, CUMBRIA
AND
BEN TROTTER
BY E-MAIL, NO ADDRESS SUPPLIED

A The industrial process for shelling Brazil nuts involves soaking them in water for 24 hours, then immersing them in boiling water for five minutes to soften the shell. The shells can then be removed from the nut quite easily by hand and the nut will need to be dried. If a Brazil nut is shelled in this way it is important to eat it soon, as the boiling 'kills' it.

Jason Hiscock
Walton-On-Thames, Surrey

A The answer to cracking a Brazil nut, or any other hard nut, is to use a pair of Mole grips. Every engineer has a pair of these grips and they are very handy at Christmas.

After some practice adjusting the grip jaws, only the minimum force need be applied to split the shell without bursting it. Cracking similar sized nuts in batches, such as a bag of Brazils or a handful of walnuts, makes the process faster.

Because of their triangular cross section, they may need to be turned by 120 degrees and have similar pressure applied in another direction.

Richard Hames
Datchet, Berkshire

A The cross section of a Brazil nut is roughly an equilateral triangle, with a slightly shorter base. But if you look inside the nut, you will find that while the shell is usually attached across this base, it tends to be free nearer the apex.

Most people try to crack the nut the easy way— by squeezing the crackers across the base, which is the shortest side. This transfers all the force directly to the nut itself—hence the shattered mess. Instead, apply the crackers across one of the sides, holding the nut between the apex and one of the lower edges. In most cases, the side wall will split neatly, leaving the nut whole and unshattered. If the other side wall is broken the same way, you will get more whole nuts than broken ones.

Doug Cross
Honiton, Devon

In Australia we grow the extremely hard macadamia nuts.

In Queensland, at an orchard where these nuts are grown, you can buy a hand-held vice of simple design, which cracks the nuts gradually as it is tightened.

RICHARD SELIGMAN
MELBOURNE

The whole Brazil nuts are placed into industrial pressure cookers called autoclaves and boiled until the shell softens slightly and expands, leaving the nut loose inside. Today most are machine cracked. They are fed between steel rollers with a gap slightly smaller than the nut itself and set at tolerances so just enough pressure is applied to crack the shell. The unscathed kernel then falls free. Smaller, older companies may crack their nuts using a vice-like implement, but this is not as sensitive and results in more wastage—Ed, (with thanks to chocolate makers Thornton).

DIRTY BUSINESS

In James Bond films, a gun with a silencer is used to dispose of bad (and good) guys. How does the silencer work?
JEREMY CHARLES
CHESHAM, BUCKINGHAMSHIRE

Silencers are more properly called sound moderators or suppressors and are widely used by hunters to reduce noise levels from the discharge of firearms, particularly sporting rifles and air weapons. A sound moderator is essentially no more than a series of baffles coupled to an expansion chamber, contained within a tubular attachment which screws on to the end of the firearm's barrel.

The noise of the discharge of most firearms is made up of two components. The first comes from the rapid expansion of propellant gases as they leave the muzzle. The second is the supersonic crack of the bullet. It is not possible to reduce the sound level of a supersonic bullet, but a sound moderator fitted to such a rifle will have some significant effect in reducing the noise signature because it controls the rate of expansion of the propelling gases.

For a sound moderator to be really effective, it must be used with ammunition whose projectiles travel at less than the speed of sound. In such cases, the noise of the discharge is greatly reduced and may not even be recognizable as a gun.

It is not possible to fit a sound moderator to a revolver because the gap between the barrel and the front of the cylinder means that about 5 per cent of the propellant gases escape, contributing to the overall noise of the discharge. Otherwise, sound moderators can be fitted to any type of firearm.

I once watched a Second World War Sten submachine gun fitted with a large, integral moderator being fired using special subsonic ammunition. The results were impressive: the only noise that came from the weapon was the clatter of its bolt.

In the public imagination, sound moderators for firearms invariably have a James Bond or underworld image. In reality, they are widely used in the countryside by hunters who wish to play their part in cutting noise pollution.

BILL HARRIMAN
THE BRITISH ASSOCIATION FOR SHOOTING AND CONSERVATION,
WREXHAM

The first successful silencers were patented in 1910 by the American inventor Hiram P. Maxim (son of Hiram S. Maxim of Maxim machine gun fame). His devices were of the baffle type, which is still in common use today. A baffle silencer typically consists of a metal cylinder, usually divided into two sections, which is fixed to the muzzle of the firearm.

The first section, which is typically about a third of the silencer's length, contains an 'expansion chamber' into which the hot gases that follow the bullet out of the muzzle can expand to dissipate some of their energy. The expansion chamber may contain a wire mesh cylinder, whose function is to break up the column of gas and to cool it by acting as a heat sink.

The second section consists of a series of metal baffles, with a central hole to allow the passage of the bullet. The function of the baffles is to progressively deflect and slow the flow of gas emerging from the expansion chamber, so that by the time the gases emerge from the silencer, their flow is cooler, at low velocity, and silenced. A motorbike silencer works on exactly the same principle.

There are also variations on this theme: some designs consist entirely of baffles, while others are based entirely on one large expansion chamber. In fact, a plastic soft drinks bottle can be made into a fairly efficient silencer that will work for a limited number of shots before it breaks up.

Silencers usually work best with cartridges that fire subsonic ammunition, because this eliminates the sonic crack which is produced by a bullet that goes faster than the speed of sound.

Some silencer designs slow the bullet to subsonic speed by means of ports cut into the barrel, with the ported section extending to protrude into the expansion chamber. These ports bleed off gas from behind the bullet, thereby reducing bore pressure and, eventually, the velocity of the bullet. In other designs, the baffles are made from an elastic material with a central hole smaller than the bullet. These 'wipes' are pushed open by the passage of the bullet and close when it is past. The idea is that they further slow the exit of gas. Not surprisingly, the wipes can wear out rather quickly and can affect the accuracy of the bullet.

A second, but less common, type of silencer is the 'wire mesh' design. These usually have the same expansion chamber as the baffle type, but the baffles are replaced by a column of knitted wire mesh with a central hole for the bullet. Here, the wire mesh acts to disrupt the column of gas as in the baffle design, while at the same time acting as a heat sink to cool the hot gas and hence quieten it. Criminals have been known to improvise this type of silencer, using wire wool or steel pan scourers to form the mesh.

The very latest innovation in muzzle-mounted silencers is the so-called 'wet' silencer (or 'wet can' in the US). These designs allow the use of water or a lubricating oil. On firing, the hot expanding gases are cooled, and therefore quietened, by the exchange of heat into the liquid.

Wet silencers allow the designer to produce much smaller or quieter designs.

An alternative approach to silencer design which dispenses entirely with the muzzle-mounted silencer has appeared from Russia. Instead it uses a special cartridge in which the bullet is pushed out by a propellant-driven piston. The piston is stopped by the neck of the cartridge, trapping the hot, noisy gas entirely within the chamber of the firearm.

It is fair to say that Hollywood takes great artistic liberties with silencers. Most real designs are very much larger than the cigar-tube sized ones typically shown on film and usually much less simple to fit and remove. Despite what is shown in films, it is usually impossible to silence a revolver because the gap between the cylinder and the barrel allows gas to escape.

Finally, forget the distinctive 'phut' produced by James Bond's silencer. Real designs are more likely to produce a muffled crack, or to sound like a car door being slammed.

Hugh Bellars
By E-Mail, No Address Supplied

IT'S THE PITS

Q *How does the service fault machine at Wimbledon and other major tennis tournaments work? It detects when a ball is out of play but doesn't seem to react to players' feet.*
Ian Hunter
Oxford

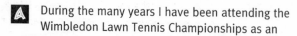

A During the many years I have been attending the Wimbledon Lawn Tennis Championships as an honorary steward I have seen the gradual incorporation of this gadget, from the time when the inventor himself came onto the sacred grass of the Centre Court to set it up, to its everyday use. It is just an infrared beam sent from a generator on one side of the court to a receiver on the other. If the beam is cut by a ball then it emits the familiar beep. The setting is crucial, because the ball is deemed in the court if it touches the line, so this must be taken into account when the beam is positioned, which consequently must be past the service line.

Yes, your choice of title was very apposite. Former champion John McEnroe frequently complained about it and, as is permitted, many times asked for it to be turned off. The latest bit of technology is the electronic net cord judge, which issues a beep when the ball touches the net. I have never heard this questioned, and the familiar figure of the human net cord judge with an ear pressed to the net support is disappearing quickly.

LAURIE NORTH
LONDON

KLINGOFF

Q *Why doesn't cling film cling to a metal bowl as well as it does to an equally smooth glass or ceramic one?*
TIM BLOOMFIELD
LETCHWORTH, HERTFORDSHIRE

A Cling film, known as cling wrap in the US, works because it acquires an electric charge as it is peeled from the roll. It can then stick to an insulating body by the same mechanism that an uncharged piece of paper sticks to the charged glass of your computer or television screen.

The mechanism relies upon the cling film and the object to which it is sticking being at a substantially different electrical potential. This works when the object is an insulator. When the object is metal, the charge on the film is dissipated throughout the object, so negating the effect.

Old cling film taken off the roll doesn't work either. After a while, the charge breaks away, and the clinginess is lost.

ALISTAIR HAMILTON
BY E-MAIL, NO ADDRESS SUPPLIED

A Cling film becomes charged with static electricity as it peels from the roll. You can sense the charge by peeling some off and holding it near your face—you will feel the hairs on your cheek stand up. Metal drains away static—glass (or plastic) retains static on its surface. The more static, the greater the cling.

JEFFREY WELLS
BY E-MAIL, NO ADDRESS SUPPLIED

NO FLAKES

Q *How does antidandruff shampoo work?*
EUGENE
BY E-MAIL, NO ADDRESS SUPPLIED

A Dandruff is thought to be caused by overgrowth of yeasts such as *Pityrosporum ovale* which live on normal skin. This overgrowth causes local irritation resulting in hyperproliferation of the cells (keratinocytes) forming the outer layer of the skin. These form scales which accumulate and are shed as dandruff flakes.

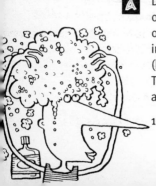

Antidandruff shampoos work by three mechanisms. Ingredients such as coal tar are antikeratostatic and they inhibit keratinocyte cell division. Detergents in the shampoo are keratolytic—they break up accumulation of scale. Finally, antifungal agents such as ketoconazole inhibit growth of the yeast itself. Other components such as selenium sulphide also inhibit yeast growth and therefore scaling.

RODDIE MCKENZIE
UNIVERSITY OF EDINBURGH

THE SPINNERS

Q *Why does a Frisbee need to spin in order to fly?*
LAURIE POST
OXFORD

A When a rock is thrown through the air, one would not consider the resultant motion as 'flying'. This is because, first, the aerodynamic lift acting on the rock is small compared to its weight, and second, the rock will more than likely tumble as it goes. To fly properly, an object must be able to produce a lift force at least equal to its weight (without causing too much drag) and be able to maintain an orderly attitude as it passes through the air.

The first of these two criteria is quite easy to fulfil. This is essentially what wings are for. The second criterion, which is related to aerodynamic stability and trim, is much harder to fulfil, and usually requires the addition of extra horizontal and vertical wings at the rear of an aircraft, to make things work properly. This is where a Frisbee runs into trouble: it is essentially a single, round wing with no extra surfaces for trim and stability.

If a stationary (non-rotating) Frisbee is tested in a wind tunnel, it is easy to show that it can produce more than enough lift to support its weight, thus fulfilling the first criterion of flying. However, it is also apparent that the Frisbee is very unstable in pitch. That is, if left to its own devices, it would soon end up tumbling over itself, and hence fails the second criterion of flying. A Frisbee has to spin in order to fly properly because additional pitch stability (strictly speaking, pitch stiffness) is required to overcome the inherent aerodynamic instability of the Frisbee's shape . . .

BILL CROWTHER
BATH, AVON

 . . . The theory of spinning tops shows that a spinning object is usually unstable unless it is spinning about what is called its major principal axis. If it does so with enough spin then its axis of rotation stays the same, even it there are disturbing forces like air resistance.

The faster the spin, the more it can withstand disturbances. This is the principle by which gyroscopes work. The major principal axis of a Frisbee is the axis through its centre, perpendicular to its face. So if the Frisbee is spun in the normal way with enough force, its axis of spin will remain approximately constant throughout its flight. This also explains why the Frisbee flies along a plane even when it is thrown at an angle to the horizontal. Its axis, which is at an angle to the horizontal, stays constant. Moreover, Frisbees which have most of their mass at the rim work even better than spinning tops—they require less spin to achieve the same results.

JOSEPH MUSCAT
PRINCETON UNIVERSITY, NEW JERSEY

THROUGH THE HOLE

Q *I recently did a parachute jump for charity and the one disconcerting thing about the jump (apart from a fear of heights) was the large hole at the top of the parachute. Why is it there? Does it help in any way to reduce the drag on the chute?*
SUZY KLEIN
LONDON

A In the days before the apex vent (the disconcerting hole in the top of the parachute canopy), the only way that the air trapped underneath the parachute could escape was to spill out from one edge of the canopy, thereby tilting it and throwing the hapless parachutist to one side.

As the canopy swung back, more air would spill out from the opposite side, setting up a regular, pendulum-like oscillation (watch any footage of Second World War parachutists and you will see this).

As you can imagine, hitting the ground during a downswing was understandably hazardous, especially if it was also a windy day. The apex vent, by allowing the air to leak slowly out of the top of the parachute canopy, prevents this wild oscillation and makes for much safer landings.

Another benefit of the apex vent is that it slows down the opening of the parachute. Without the vent, air inflates the canopy much more abruptly, and it can damage the parachute or bring tears to the eyes of (particularly male) jumpers.
PAUL DEAR
CAMBRIDGE

WHAT'S THE RUB?

Q *How does a pencil rubber work? Do all types—conventional, plastic and putty—work in the same way?*
HELEN GEAKE
YORK

A A pencil mark actually consists of graphite particles abraded from the pencil point by the paper. These particles, which have an angular, gritty look under the microscope are, for an HB lead pencil, typically between 2 and 10 micrometres in diameter. The particles lie slightly below the surface of the paper, interlocked between its fibres. A single rub using a rubber sufficiently soft to reach between the fibres will pick up most of them. Inspection of the rubber shows the undamaged particles adhering to the surface. An effective erasing material is also abraded by the paper surface, producing the familiar small spindles of rubber or eraser material, which wrap up the graphite particles. At 200 times magnification, these look like roly-poly puddings studded with graphite raisins.
JOHN ROWLAND
DERBY

ZEUS JUICE

Q *What chemicals or substances are used to keep the Olympic flame burning on its way to the Olympic stadium?*
JAHAN ANZSAR
LONDON

The Olympic torches used in the 1996 Summer Olympics held here in Atlanta used propane gas stored in canisters in the handle.
TERRY BLANTON
ATLANTA, GEORGIA

BUBBLES, LIQUIDS, AND ICE

AIM AND POUR

Q *When I open a carton of milk I have to pour the liquid quickly from the opening so that it goes into my glass. If I tip the carton too slowly, the milk runs down the underside of the carton and pours onto my foot or the floor. Orange juice and other liquids do the same. Why do they stick to the carton when poured slowly?*
Tom Khan
Bradford, West Yorkshire

A When a carton of liquid is tipped during pouring, the free surface of the liquid in the container is raised relative to the opening. This creates a pressure difference between the free surface and the opening, which forces fluid from the carton. In addition to this pressure force, there are also surface tension forces acting on the fluid that tend to draw the fluid towards the surfaces of the container. At high pouring speeds, the pressure force is much greater than the surface tension forces, and the fluid will leave the carton in an orderly fashion, following a predictably curved (parabolic) path towards a glass below.

However, at low pouring speeds, a point is reached where the surface tension forces are sufficient to divert the path of the fluid jet so that it fails to leave the opening cleanly and becomes attached to the top face of the carton (assuming a flat-topped carton). Once attached to a surface, a jet of liquid will tend to remain attached to that surface due to these surface tension forces and a phenomenon known as the Coanda effect. This occurs when a fluid jet on a convex surface (such as a water jet from a tap curving round the back of a spoon) generates internal pressure forces that effectively suck the jet towards the surface.

The combined result of surface tension and the Coanda effect enable an errant flow of fluid to negotiate the bend from the top face of the carton round to the carton's side, thus ensuring maximum transport of fluid from the carton to your shoes.

Experiments have shown that when cartons are full, the 'glugging' that occurs as air is sucked in to replace the lost fluid causes the fluid jet to oscillate, leading to periodic surface attachment of the jet (and wet shoes) even at relatively high pouring speeds.

BILL CROWTHER
AEROSPACE DIVISION, SCHOOL OF ENGINEERING, UNIVERSITY OF MANCHESTER

The Coanda effect or 'wall attachment' is named after the Romanian Henri Coanda (1886–1972) who invented a jet aircraft propelled by two combustion chambers, one on either side of the fuselage pointing backwards, and situated towards the front of the aircraft. To his horror, on take-off the jets of flame, instead of remaining straight, clung to the sides of the fuselage all the way to the tail. At least his name has now been immortalized, thanks to this effect.

Some 30 years ago this 'wall attachment' phenomenon was used in machine control systems known as fluidics, where a small jet of fluid was used to persuade the main flow to leave the 'wall' to which it was attached and divert to another course. It then became attached to this.

JOHN WORTHINGTON
STOURBRIDGE, WEST MIDLANDS

A picture of the Coanda, the first true jet aircraft, built in 1910, can be found at

www.allstar.fiu.edu/aero/coanda.htm. The next answer describes a simple demonstration of the effect—Ed.

A The effect is seen as a general tendency for fluid flows to wrap around surfaces. An entertaining experiment consists of taking a vertical cylinder (a washing-up liquid bottle or a wine bottle) and placing a lighted candle on the far side. When you blow against the bottle, the candle is blown out, because the current of air wraps around it.
RICHARD HANN
IPSWICH

CLEAR THINKING

Q *My nine-year-old son has come up with a question that has beaten my years of science education. Why can you see through water? Can anybody help us understand what it is about water—or any transparent medium— that lets light through it?*
MARTIN J. BENWELL
CHARTERHOUSE COLLEGE OF RADIOGRAPHY,
LONDON

A The question should be turned around the other way. What matters is not 'what lets light through' transparent media, but what stops it in opaque ones.
HUGH DUKES
LUTON, BEDFORDSHIRE

A Light is electromagnetic radiation which exists as packets of energy called photons. These act like waves and the energy of a photon is precisely related to its wavelength.

Solid objects, or matter, consist of particles (atoms and molecules) which include electrons. Electrons only exist at certain energy states and a photon which makes contact with a particle may excite an electron or molecule to a higher energy state. This can only occur if the energy of the photon is exactly the same as the energy required to raise the electron from one energy state to another. If this occurs, photons of this particular wavelength are absorbed and the matter is opaque to the light radiation.

Some substances are translucent. Photons can pass through matter if they have wavelengths which are not absorbed and many materials allow photons to pass, but some deflect the photons in many directions (scattering). Such substances behave as they do because waves can interact with objects which have a similar size to their wavelength. This is known as interference. Crystals are good examples of translucent substances because they are regular arrays of atoms, and the distances between their atoms can be similar to the wavelengths of the photons. So photons which pass through crystals can be deflected (refraction) and are scattered.

A material is transparent if it transmits photons which continue in their original direction, or are all deflected by the same change in direction. Such transmitted radiation retains any image which it had when it entered the material. Transparent materials contain very little crystalline material and their atoms and molecules are arranged in an irregular pattern making them glassy or amorphous. This induces negligible scattering because the net effect of any scattering that is induced is zero.

Materials may be transparent to some wavelengths of light but not others. Visible light has a limited range of wavelengths. Even so, different wavelengths are seen as different colours. Some materials may be transparent to some colours but opaque to others which explains, for example, tinted spectacles.

RICHARD COURTNEY
CHELTENHAM, GLOUCESTERSHIRE

NATURAL GUNK

Q *When walking in the Pennines I often notice that streams have lumps of foam floating on their surface, particularly near waterfalls. There are no sources of human pollution upstream, so what causes the foam?*
PENNY JOHNSON
BISHOP'S STORTFORD, HERTFORDSHIRE

A The foam is created by natural surfactants released into the water from decaying plant material upstream. Perhaps the most familiar surfactants are soap and washing-up liquid, but there are numerous others, and they are not all manufactured.

All surfactant molecules have a polar region which is attracted to water molecules (it is hydrophilic) and a nonpolar region which is repelled by water (hydrophobic). Organic breakdown products such as humic acids consist of large molecules that may have several hydrophobic and hydrophilic regions. Hence they can act as natural surfactants.

In your Pennine waterfall air is mixed into the water. Surfactants are concentrated at the point where air meets water, surrounding each air bubble with their hydrophilic regions in the water and their hydrophobic regions in the air. As these

BUBBLES, LIQUIDS, AND ICE 175

bubbles rise, the 'skin' of surfactants follows. In fact, bubbling is sometimes used as a method of removing dissolved organic matter from water.

Water has a high surface tension, so bubbles rising in pure water burst immediately once they reach the surface because the thin layer of water surrounding them pulls together into a single water drop. However, surfactants lower this surface tension and make the bubbles last much longer—hence the foam.

You can see a similar phenomenon along the seashore, especially after autumn storms. Organic molecules are released into the seawater by microscopic algae called phytoplankton when they are under stress or as they decompose after death. These surfactants, a complex mixture of oils, sugars and proteins, are stripped from the seawater by the bubbles created by breaking waves along the shore. They can form very long-lasting foam, which is washed or blown ashore to lie on the beach like a huge blanket of whipped egg white (which is, of course, another natural surfactant).

Val Byfield
Salisbury, Wiltshire

PADDLE PUZZLE

Q *Glenbrook Infants School went to the seaside for our summer outing. We had a nice time, but please can you tell me why the sea is salty. My mum doesn't know.*
John Connolly
London

A The sea is salty because the rivers that flow into it wash salts and other minerals out of the ground.

The salts dissolve in the rivers and the rivers flow into the sea. As the Sun evaporates the water from the sea to make clouds, it leaves the salts and minerals behind, so the sea is saltier than rivers and lakes.

JACK CAVE-LYNCH (AGE 9)
WELLINGTON, NEW ZEALAND

John Connolly is a brainy guy
Asking questions and wondering why
The salty sea which is such fun
When splashing in the waves and sun
Is not freshwater from the tap
Or from a bottle with a cap;
So he will learn that salt and sea
Mix just like sugar into tea
And that many other kinds of salt
Dissolve into this briny malt,
Sodium chloride, the salt of table
Has other friends within its stable
Potassium Ch, magnesium Ch, and iodide
All flow solvent with the tide.
So now, dear John, you clever lad
Off you go—tell mum and dad!

RAY HEATON
SOLIHULL, WEST MIDLANDS

POOL PULL

The density of saltwater is higher than that of freshwater, so swimming in saltwater should be marginally faster than in freshwater because swimmers are carried slightly higher in the water and suffer less friction.

Given the split-second timing of world records, are the water densities of the various competition pools measured and controlled

in the same way that wind speeds for athletic track events are kept within limits?
ROTTERDAM, THE NETHERLANDS

 It is a misconception that the higher you are in the water, the faster you can swim. On the contrary, the highest speeds are reached by swimming completely submerged, because this allows for more efficient transfer of momentum to the water (which creates forward thrust according to Newton's third law), and because less energy is wasted splashing water.
HANS STARNBERG
DEPARTMENT OF PHYSICS
GOTHENBURG UNIVERSITY, SWEDEN

 Turbulence at the surface of the water increases drag and slows swimmers down. Trained swimmers know that swimming underwater is faster than swimming at the surface.

In all the different types of competition stroke, from Olympic level down to club level, and at all distances, the number of underwater strokes is strictly limited (especially at the turn) for this very reason. Competitors are disqualified and records annulled if they break the rules.
KEVIN DIXON-JACKSON
ECCLES, LANCASHIRE

The rules on turning vary depending on the stroke. In freestyle, backstroke and butterfly, competitors may travel no more than 15 metres underwater after a turn. The rules for breaststroke are more complicated, but in effect swimmers are allowed only one full kick and one full arm pull before surfacing.

178 BUBBLES, LIQUIDS, AND ICE

The density of pool water in competitions is not regulated, but the temperature of the pool must be kept at 26 °C, although a variance of 1 degree above and below this temperature is permissible in international competition—Ed.

SPACE HOP

Q *Imagine you're sitting in a pub with a pint of real ale. Gravity keeps the beer in the glass and the bubbles rise through it. Now imagine that you have a globule of beer floating in a spacecraft in zero gravity. What happens to the bubbles? What direction do they move in—if they move at all? Are they the same size as on Earth? Would the beer have a frothy head? Are there likely to be any other unusual effects?*

STEPHEN STEWART
BY E-MAIL, NO ADDRESS SUPPLIED

A In microgravity, surface tension tends to be the driving force behind fluid behaviour. Once released from whatever container it was in, your blob of beer would just float there. However, if you opened a can of beer in orbit, you'd create a nifty little beer cannon that would cover the wall with several globs of beer.

Bubbles would still form in your beer globule, because the carbon dioxide would still come out of solution under room temperature and pressure, but they wouldn't move in any direction. Not only that, but the larger bubbles (and head) in Earth beer form because the bubbles float to the top of the glass and bump into each other on the way. Space beer would have a number of bubbles throughout, so you'd just get a foamy mass.

Astronauts aren't allowed to drink carbonated drinks in orbit, because the body relies on gravity to burp excess gas. No beer is one of the many sacrifices one must make for space exploration.

TODD DARK-FOX
THOUSAND OAKS, CALIFORNIA

 Bubbles are likely to be fewer and larger in microgravity because as they form, they remain at the nucleation sites instead of drifting off. However, their growth may be slower because it would depend more on diffusion through the liquid and less on circulation. On Earth, the behaviour of beer depends largely on gravity. In free fall, surface tension, momentum, vapour pressure and diffusion dominate, so small bubbles are less likely to meet and fuse, but big bubbles are less likely to reach the surface and burst.

To get a head on top of a zero-gravity slug of beer you first must define where the top is. One way is to catch the slug in a container and substitute for gravity by swinging it round and round, amusing your colleagues by smashing their instruments and splashing them with beer. More elegantly, you can husband a small slug until it is nicely spherical, then gently blow on it to start it rotating. Rotation defines 'the top' as being inside the slug, and the scattered bubbles will congregate along the axis of rotation.

Forces hardly noticeable on Earth have weird effects in microgravity and complicate working in space. Localized drying and the lack of convection cause differences in viscosity, vapour pressure and surface tension, which make fluids creep, drift and distort unexpectedly.

JON RICHFIELD
DENNESIG, SOUTH AFRICA

WAVE POWER

Q *What mechanism transforms gusting wind energy into the regular wave train of ocean swells and what determines their amplitude and frequency?*

FRANK SCAHILL
EASTONVILLE, NEW SOUTH WALES

A When the wind blows over a flat sea surface, small ripples form. These probably correspond to individual strong gusts, are disorganized and have no fixed direction or frequency.

However, as the wind continues to blow, two things happen. First, the waves interact with each other to produce longer waves which means lower frequency. Secondly, the wind pushes these larger waves and puts even more energy into them. As long as the storm lasts, the wind will make the waves larger and the wave dynamics will create longer and longer waves.

Some waves will become too steep and break but, in general, the total amount of energy will keep increasing. These locally generated waves are known as 'wind-sea'. Their energy depends on how long the wind has been blowing (its duration) and over what distance (the fetch). The waves on the sea surface are not a simple wave train but a complicated random surface.

It is impossible to give a simple amplitude and frequency for a system as complex as this. Instead, significant wave height, the mean height of the highest third of the waves, is used to describe how large the waves are, and the peak period, the time between the dominant or most energetic waves, is used as a measure of frequency. On average, there will be a wave twice the significant wave height every three hours.

Eventually, the energy put into the sea by the wind will be balanced by the loss of energy, mainly through waves breaking. At this point, the waves will cease to grow and the sea is described as 'fully developed'. In a wind of 20 metres per second (a Force 8 gale), a fully developed sea would have a significant wave height of 9 metres and a peak period of 15 seconds.

Waves can travel thousands of kilometres from the point of generation. Unlike light or sound waves, as sea waves become longer (and the frequency gets smaller), they also travel faster.

Waves that escape from the storm which generated them are known as 'swell'. They have a much narrower range of periods and are almost regular wave trains. Because no more energy is put into them, none is dissipated by breaking, and they continue across the ocean until they hit the land.

Because different frequencies travel at different speeds, as swell travels across the ocean it separates into its individual components. So the significant wave height and peak period of the swell are set by the wind speed, duration and fetch from the storm that generated them.

PETER CHALLENOR
SOUTHAMPTON OCEANOGRAPHY CENTRE

A Wind energy first gives rise to a wind-sea. Waves in a wind-sea are steeper and more chaotic than swell, and are accompanied by 'whitecaps', the breaking crests of waves. The longer the wind blows, the longer the wavelength of the predominant waves in the wind-sea.

When the wind ceases or the wind-sea waves move out of the generating region, whitecapping continues for a time and is accompanied by a lengthening of the waves, until they are no longer steep enough to sustain whitecaps. The wind-sea then becomes swell.

Surface waves on liquids are dispersive, which means that different wavelengths travel with different velocities. The longer wavelength swell travels faster and arrives at the observer first.

With the passage of time, the swell wavelength becomes shorter as shorter, slower wavelengths arrive. Swell from a storm that formed thousands of kilometres away may persist for several days, steadily getting shorter because of its dispersion.

Dispersion acts as a filter, so only swell within a narrow bandwidth is present in one region of ocean at any time. This is why swell looks so uniform when viewed from an aircraft.

Generally, swell decreases in amplitude as it travels away from the source region because its energy is spread over an ever larger region of ocean.

However, this is not the whole story. A following wind will generate a wind-sea that can transfer some of its energy to the swell and increase the amplitude of the swell without changing its wavelength. Likewise, an opposing wind-sea can diminish a swell.

John Reid
Formerly of Hobart Laboratories of the Division of Marine Research, Tasmania

THE WET STUFF

Why do porous objects, such as fabrics, paper and concrete appear darker when they are wet?
Natalie Bleicher
East Barnet, Hertfordshire

Porous objects become darker when they are wet because the many tiny reflecting surfaces that cover their surface become filled in by the water and cease to reflect specular light back to the observer. This makes the object appear darker.
JAMES BAILEY
SUSSEX, WISCONSIN

YOUR BODY

BRAIN WAVES

Q **_Why are there fissures or folds in the surface of the brain?_**
BRIAN LASSEN
CANBERRA, ACT

A The brain has fissures to increase the surface area for the cortex. Dimmer animals such as rats have smooth brains.

Much of the work carried out in the brain is performed by the top few layers of cells—a lot of the brain's volume is, in effect, point-to-point wiring.

So, if you need to do lots of processing, it is much more efficient to grow fissures than it is to expand the surface area of the brain by increasing the skull diameter.

ANTHONY STAINES
BY E-MAIL, NO ADDRESS SUPPLIED

A Evidently they are there to maximize the surface area of the brain cortex. The real question is why this is necessary.

The answer probably lies in the relative number of short-range and long-range connections needed.

If many short-range connections are required, it makes more sense to pack the processing units into thin, almost two-dimensional, plate and reserve a third dimension for long-range connections.

If the neurons were distributed homogeneously over the whole volume of the brain, long-range connections would possibly be shorter, but they would take up the space between the computational units of the brain and thus lengthen the short-range connections, increasing the overall brain volume.

JANNE SINKKONEN
FINLAND

Another possible answer lies in the amount of heat produced in the brain—Ed.

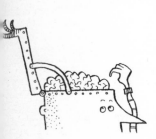

A Brain tissues consume massive amounts of energy and the resulting heat that is generated has to be dumped. Put your hand on your head and feel how hot it feels compared to your thigh.

Brains of lower vertebrate animals lack extensive folds because they have relatively less heat to get rid of.

Humans, on the other hand, have large brains which do a lot of work. The extra folds in our brains increase the surface area for blood vessels to dump the excess heat produced by all that hard thinking. If our brains were to evolve into more complex and larger organs, their folding would have to increase exponentially in order to be able to release the additional heat that they would produce.

GERALD LEGG
BRIGHTON, SUSSEX

A Many intelligent vertebrates are endowed with both large brains and a very convoluted cerebral cortex. Therefore, although the dolphin and the shark are of similar size, the dolphin's brain is considerably larger and more convoluted than that of the shark.

The cat and the rabbit are also of similar size to each other but the cat, being carnivorous, has a more complex lifestyle, presumably necessitating greater intelligence, so the cat has a convoluted brain while the rabbit does not.

The size of the animal is also an important factor. Mice and rats, while showing intelligent behaviour, have hardly any fissures in their brains but elephants and whales have brains that are even more convoluted than a human's.

It is interesting that this larger amount of cerebral cortex does not necessarily correspond to a larger number of cortical nerve cells. It turns out that these are larger and more widely spaced in large animals.

One reason for this is that the ratio of glia to neurons is considerably greater in these large vertebrates. As a result, the cerebral cortex—a laminar structure—needs to become folded to contain the number of neurons that smaller animals can afford to have in a non-folded cortex.

E. RAMON MOLINER
NORTH HATLEY, QUEBEC

CONCENTRATION

Q **People doing a tricky job will stick their tongue out and clamp it between their lips. Why? Does this happen in all cultures?**
STEVE TOWNSEND
NO ADDRESS SUPPLIED

A When you need to concentrate on something, say a word problem, you are using the hemisphere of the brain also used for processing motor input. It is amusing to see people slow down when they are thinking of something difficult while walking. This is caused by interference from the two activities fighting for the same bit of brain

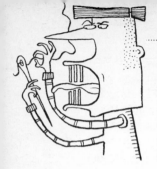

to process them. I suppose by biting your lip or sticking your tongue out, you are suspending motor activity and also keeping your head rigid, to minimize movement, and hence interference.

MELANIE WESTERN
BY E-MAIL, NO ADDRESS SUPPLIED

A Large areas of the brain are devoted to control of the tongue and to the receipt of sensation from it.

Perhaps with the tongue held rigid against the teeth or lips, the activity of those areas is subdued, allowing delicate tasks like threading a needle to proceed with less interference.

BARRY LORD
ROCHDALE, LANCASHIRE

IT'S NOT

Q *Sorry, but I had to ask, why is nasal mucus often green?*

A Of all the body cavities in contact with the outside world, the nose is probably one of the most hospitable: it is warm, very well-aerated, moist, and supplies unlimited quantities of bacterial food secreted continuously by the nasal mucosa (mucus contains quantities of glycoprotein and dissolved salts). In other words it is an ideal breeding ground for bacteria, which are always present.

Many of the common bacteria associated with humans are coloured, *Staphylococcus aureus* is a golden yellow, for example, and *Pseudomonas pyocyanea* (to give it its older, but more explicit name) is a shade of blue.

Normally these and the multitude of other organisms that are inhaled continuously into the nose are flushed out by runny mucus, which is swallowed. The bacteria are usually digested.

However, if a situation arises where the flow of mucus slows down and then becomes much thicker in response to an infection of any kind, then the bacteria, in their ideal home, can multiply and produce the coloured mucus described. This, as many parents know, is one of the less endearing characteristics of babies and young children!

And, by the way, if you're still wondering where the green colour comes from, remember what happens when you add blue to yellow.

LAURIE NORTH
LONDON

 Your previous correspondent suggested that the green colour is caused by a combination of golden-yellow *Staphylococcus aureus* and blue *Pseudomonas pyocyanea*. This is a rather tenacious belief. While the eighth edition of *Bergey's Manual of Determinative Bacteriology* (Williams & Wilkins, Baltimore, 1974, p. 222), still held *P. pyocyanea* 'commonly isolated from wound, burn and urinary tract infections' to be the causative agent of 'blue pus', the cause of the green colour of pus or nasal mucus is more general.

Green pus (or green nasal mucus) is caused by iron-containing myelo-peroxidases and other oxidases and peroxidases used by polymorphonuclear (PMN) granulocytes (neutrophils). These short-lived phyagoytizing leucocytes avidly ingest all sorts of bacteria and inactivate them by oxidative processes, involving the iron-containing enzymes above. The resulting breakdown product (comprising dead PMNs, digested bacteria and used enzymes), pus, contains significant amounts of iron, which gives it its greenish colour.

C. J. Van Oss
Department of Microbiology
State University of New York,
Buffalo
and
J. O. Naim
Department of Surgery
Rochester General Hospital,
New York

A Nasal mucus isn't always green. Nasal mucus produced at the beginning of a cold is clear and is produced in response to tissue damage caused by the invading rhinovirus. It only turns green a few days into the infection as neutrophils respond to clear away the cellular debris and secondary bacterial infection sets in.

Juli Warder
Abingdon, Oxfordshire

A Polymorphonuclear leucocytes are equipped with a number of enzymes, the most potent of which is peroxidase. This same peroxidase is also found in horseradish, giving it a distinctive green colour and a sharp (if fleeting) bite, as anyone who has tried Japanese wasabi paste can confirm. English horseradish sauce loses its colour due to oxidation of this labile enzyme on exposure to air. However, authentic wasabi is always served fresh.

Sorry if this response puts some readers off their sushi.

Steve Flecknoe-Brown
Sydney, New South Wales

OUCH!

Q *What causes the pain induced by a piece of silver foil touching a tooth filling?*
SIMON ODDY
MELKSHAM, WILTSHIRE

A The questioner is inadvertently replicating a famous experiment first performed by Luigi Galvani in 1762.

When two dissimilar metals are separated by a conducting liquid, a current will flow between them, and this current can be used to stimulate nerves.

This is exactly what happens when silver foil appears to touch the amalgam of a filling. A thin film of saliva actually separates foil from filling, and because saliva is a reasonable electrolyte, containing various salts, a current will flow between the tooth and the filling. As the filling is close to the dental nerve, the current will stimulate it, causing pain.

Galvani carried out his experiment with frogs' legs and metal probes, but the effect was the same—they twitched!

CHRIS QUINN
WIDNES, CHESHIRE

SPEAKER'S THROAT

Q *What are the bodily changes that cause us to have a dry throat when we are nervous?*
HOWARD FOSS
HONITON, DEVON

You get a dry mouth during public speaking because when you are nervous the body is set into the 'fight or flight' state. This is caused by an activation of the autonomic nervous system. It is seen throughout the animal kingdom, and has evolved to help the animal deal with dangerous situations—when escaping from predators, for example.

The nerves are selectively activated, depending on how important they are for the response. Because eating is not considered to be important at this time—you want to get the hell out of the place—the nerves to your mouth that control the salivary glands are suppressed, so your mouth dries up. In addition, your pupils dilate and the blood vessels to your muscles and heart are enlarged in order to get the blood to the most important organs needed for whatever drastic action is necessary.

M. Scotten
By E-Mail, No Address Supplied

This is tied to the 'fight or flight' reaction. In a tense or dangerous situation, your body shuts down all unnecessary functions, including the digestive system. Your saliva glands are a part of it. You don't need to digest your last meal if a lion is trying to make you his next. This is also where butterflies in the stomach come from.

Bill Isaacson
By E-Mail, No Address Supplied

SPOT LUCK

Q **Does eating sweet things really cause spots?**
Andrea Walker
London

This is a widely held myth but I think it serves a purpose for parents. They can tell their children that if they eat sweets they will get spots—even though it isn't true, parents find it useful to perpetuate the myth. Spots or acne are caused by the over sensitivity of the skin to the male hormone testosterone. This causes increased oil production in the skin and a growth change in the follicular keratinocytes.

The result may be a partial blockage of a pore, either at its opening or inside it. The accumulation of oil leads to the proliferation of the bacterium *Propionibacterium acnes*, which in turn causes inflammation and the red, pussy spot that is the hallmark of acne.

Extensive studies in the US have examined the relationship between diet and acne, including one controlled trial where some patients ate large quantities of chocolate and other sweet foods while others avoided such foods. The number of spots that developed on each person was counted. There was no significant evidence that sweets, chocolates, fried or fatty food had any impact on acne.

As with any medical condition, there were always exceptions to the rule. I certainly have patients who tell me that if they even look at a sweet, they will develop another spot. Obviously we take such claims seriously, and we do tell these patients to avoid sweets. However, in about 95 per cent of people who suffer from acne, it is certain that diet really plays no part at all.

Tony Chu
Dermatology Department, Imperial College, London

WHAT'S THE CRACK?

Q *What causes the noise when you crack your knuckles or any other joint?*
MARTY BROWN
BY E-MAIL, NO ADDRESS SUPPLIED

A A click or crack is often heard when a joint is moved or stretched. When the pressure of the synovial fluid in the joint cavity is reduced, this may create a gas bubble and generate a popping sound. The sound may also be the result of separating the joint's surfaces, which releases the vacuum seal of the joint.

These noises are sometimes produced during osteopathic treatment, but this does not prove that the treatment has worked, nor does their absence mean the treatment has failed. The test of success is whether the joint's range and ease of movement have been improved.
WILL PODMORE
THE BRITISH SCHOOL OF OSTEOPATHY, LONDON

A All the soft tissues of the body, including the capsules of joints, contain dissolved nitrogen. When a vacuum is applied to the joint space by pulling on the bones, say by flexing the fingers strongly, nitrogen suddenly comes out of solution and enters the joint space with a slight popping sound.

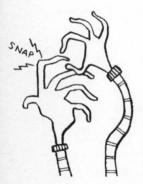

Radiologists often see a small crescent of gas between the cartilages of the shoulder joint on the chest X-rays of children who are held by the arms. This is due to the force of pulling on the arms causing nitrogen to evaporate into the joint space. It can infrequently be seen in the hip too.

Small, highly mobile bubbles sometimes appear within the hip joint of a baby being screened for congenital dislocation of the hip using ultrasound. This usually happens if the infant is struggling and has to be held firmly. The bubbles disappear after a short while when the nitrogen dissolves again.

If the fingers were X-rayed immediately after cracking the knuckles a fine lucency, as a result of thousands of tiny opaque bubbles, would probably be visible between the ends of the bones.

Tony Lamont
Mater Children's Hospital,
Brisbane, Queensland

WINE INTO WATER

Q *No matter what colour of drink one consumes, when the liquid finally leaves the body the colour has gone. What happens to it?*
P. Beeham
Witney, Oxfordshire

A The liquid that leaves the body is almost unrelated, in chemical composition, to the liquid consumed. Any substance, solid or liquid, that goes down the oesophagus, passes through the digestive tract and, if not absorbed, is incorporated into the faecal matter. Urine, in contrast, is created by the kidneys from metabolic waste produced in the tissues and transported through the bloodstream.

Any coloured compound that you drink either will or will not interact biochemically with the body's systems. If it does, this interaction (like any other chemical reaction it might undergo) will tend to alter or eliminate its colour. If it does not, the digestive system will usually decline to absorb it and it will be excreted in the faeces which, you will have noticed, show considerably more colour variation than the urine.

STEPHEN GISSELBRECHT
BOSTON, MASSACHUSETTS

 Coloured substances in food and drink are usually organic compounds that the human body has an amazing ability to metabolize, turning them into colourless carbon dioxide, water, and urea. The toughest stuff is often taken care of by the liver, which is a veritable waste incinerator. However, on the very infrequent occasion when the intake of coloured substances exceeds what the body can quickly metabolize, the colour is not necessarily removed as the liquid leaves the body. This is well known to anyone who has indulged in large quantities of borsch (Russian beetroot soup).

HANS STARNBERG
GOTHENBURG, SWEDEN

INDEX